BEEKEEPING

How to Start a Beekeeping Hobby at Low Cost

(A Complete Guide for Keeping Bees and
Harvesting Honey)

Theodore Wallace

Published by Andrew Zen

Theodore Wallace

Beekeeping: How to Start a Beekeeping Hobby at Low Cost (A Complete Guide for Keeping Bees and Harvesting Honey)

ISBN 978-1-77485-192-0

Legal & Disclaimer

The information contained in this book is not designed to replace or take the place of any form of medicine or professional medical advice. The information in this book has been provided for educational and entertainment purposes only.

The information contained in this book has been compiled from sources deemed reliable, and it is accurate to the best of the Author's knowledge; however, the Author cannot guarantee its accuracy and validity and cannot be held liable for any errors or omissions. Changes are periodically made to this book. You must consult your doctor or get professional medical advice before using any of the suggested remedies, techniques, or information in this book.

Table of Contents

Introduction

The book offers the most effective steps and strategies for how to grow honeybees in your garden and what I refer to as an DIY Beekeeping. This book was carefully made to guide your through creating an eco-friendly beekeeping program Get the top equipment for beekeeping and also other suggestions and ideas to help you maintain your beekeeping efforts for quite a long duration. The book was designed to guide you through a series of steps you can follow in constructing a hive and staying clear of some of the most severe issues that come along with beekeeping.

Here's a glimpse of what you can expect to learn;

How do you setup your hive and the best way to learn about various beehive setups,

What kind of equipment do you will need to build and maintaining your beekeeping business?

How and when to collect honey,

How to attract more bees to avoid common issues like bee stings

The types of materials required for raising bees and

How can you commercialize your beekeeping efforts.

Beekeeping is a cheap hobby that can become a business venture to earn additional streams of revenue. There are many components in beekeeping, and you have to know the various aspects, particularly if are looking to become an effective beekeeper. You should not compromise the quality of your product with price, but it is important to keep in mind that beekeeping can be relatively affordable however if you wish to expand your business and improve its viability commercially it is necessary to make more investments in capital.

Chapter 1: An Introduction To Beekeeping

If this is your first experience with keeping bees, it's likely to be asking lots of questions. Most likely, you've observed beehives on farms or fields in the past but they're usually too far away to be able to observe the bees. Perhaps you've been lucky enough to have a chance to visit family members or friends who have bees in their hives, and you were able to observe the hives in close proximity.

If you've never tried beekeeping in the past it is now the right time to become familiar with the activity before you begin.

Honey has been collected by humans since the dawn of the world. With excellent skills (and an extremely high tolerance to pain) humanity has been said to have collected honey for more than 10,000 years. But, there is evidence of Egyptians are the first to domesticate and harvested honey bees for over five thousand years.

While in the past collecting honey out of bees seemed to be a challenging task, today's beekeeping is a subject that requires thorough research and you must be aware of what you're getting yourself into prior to beginning. If the bees are not happy in their hives or feel they're in danger they'll move out and search for an alternative location to reside and leave you back to the beginning.

This being said it is crucial that you understand how to take care of your bees before you put them in the pictureso that you can be sure they are happy in the hive that you have prepared for them, and then decide to remain in the hive.

In the beginning, it's crucial to check the local ordinances in the place you liveand decide whether you require an permit to set up your hive. It's not common to require an permit to keep bees, but it can happen and you shouldn't create a hive only to lose it from you or be penalized later for not realize you required an permit.

Then, you must realize that bees follow travel patterns and they usually following the same route while they collect pollen. The brains of bees are wired to remember maps of where food source is. They return to the same spot day after day collecting nectar and pollen for them to return to the honeybee hive (and transform in honey.)

Keep this in mind when you put the hive within an area that the bees' movement patterns won't disrupt your neighbors, family members, or your pet. Bees are extremely busy and would prefer to not be involved in any activities with anyone else than people in their hive. therefore, doing this is as much of a gift to the bees as it is to you.

Do you think beekeeping is an overwhelming job?

The amount of effort you do in beekeeping can differ. This a remarkably seasonal pursuit, meaning that you'll be doing nothing during winter. However, you could discover that it takes up the majority of your time during the spring.

In the case of bees it is your responsibility to be the one to take charge of essential care, but let them take care of the rest. Of course, you're likely to be involved when it's time to collect honey, but remember that bees are extremely independent, and the less involved in the process, the more efficient.

If you want to succeed in beekeeping, a strategy is the most effective method. Know what a colony of bees is and what it does Know the extent of effort you're going to need to perform for ensuring that your bee colony is flourishing and know the amount (or what) you will take on to ensure it's a success.

Learn about the amount of honey you can get from bees Be sure to be generous with regards with their honey. Take only the amount they can spare and nothing more. There are many guidelines to adhere to, but believe me when I say once you've got your bearings and you'll be able to enjoy your hobby of beekeeping.

The three elements of diligence, time and commitment are the three main ingredients for the success of beekeeping.

Chapter 2: Types Of Bees And Colony Structure

Like any other enterprise it is essential to be prepared with all the information and knowledge that have to do with beekeeping because these are the keys to success.

Without them, it'll be similar to cooking with no knowledge of the necessary ingredients. Be aware of the basics of keeping bees, and get aware of the smallest of details.

A single of the crucial things you need to be aware of is the various kinds of bees. Bees are one of the primary "ingredients" of a bee keeping company, so they're very important.

Before you can know what to do with these bees, you need to be aware of the different types so you be able to recognize them. It is important to know that bees are available in a variety of kinds and each one plays an individual role.

There are a variety of types of bees needed for the existence in any colonies. Bees are social creatures. They live in groups and collaborate to benefit the entire colony, taking care of the young bees, and collaborating in the search in search of nectar as well as making honey.

Within honeybee "society" there are three distinct kinds of bees. There are drones, the queen, and the worker. Since honey bee colonies can last for a long time, all bees are working together to get through the winter, and the one following.

The worker bees are females with no reproductive organs developed. They complete all of the work needed to sustain the colony. A colony could contain between 2,000 to 6000 worker bees.

The Queen is fertile female. Her task is to create eggs that are used by the colony. If a queen bee dies or quits the colonies workers select a few of larvae from the worker. The larvae are given "royal jelly." The royal jelly triggers the larvae to transform into queens.

If they were not fed this diet these same larvae will be able to become workers. There is typically the queen bee in the bee hive. She is responsible for the production of chemicals known as Pheromones, which control the behavior of the different types of bees in the hive.

drones, also known as males. In the summer and spring months, a colony could be without drones or might have 500 drones. The purpose of drones is to depart the honeybees and, while in flight and mate with queens of other colonies.

Workers construct hexagonal beeswax cells, in which the queen lay her eggs. When broods (young bees) grow they move through four different stages. The stages include the egg, larva, the inactive pupa and the young adult. Each stage takes a specific amount of time to develop, based on the type of bee the egg will grow to turn into.

When the workers have left their cells they begin their cleaning. Their task is to wash cells, circulate air within the hive, beat their wings feed larvae, train to fly until they become proficient, collect pollen and nectar from bees who have hunted in search of it, and protect the entrance to the hive as well to hunt for food. The various tasks they perform depend on their the age of the worker.

In winter honey bees gather in a tightly clumped. The queen begins to lay eggs in January, in mid-nest. The end of winter is when the food reserves in a colony will be less than they were in the beginning of spring, when the nectar flow commences, the number of bees increases quickly.

It's when swarming happens. There will be many queens within a group that is crowded. Mother of daughters can take off, carrying more than half her employees with her.

They will hang on to the branches of a tree or wall , while scouts locate the ideal nesting spot. Within a few hours an ideal nesting spot is found, in which one of the new queens will be able to inherit the colony. So the life cycle starts once more.

Chapter 3: Beekeeping In The Country

In recent times the popularity of beekeeping has grown as a viable idea for rural businesses. is the beekeeping process for producing honey. Well, everyone loves honey. However, few think of it to be more than just a leisure activity with the potential to become far more.

A sustainable apiary is an art form and it's easier if you've got plenty of space. The bees will be able to be cared for in a more natural setting. It is therefore important to pick honey when it is necessary and to nurture your bees throughout the time it is.

February is the usual time when the queen is ready to begin to lay eggs. If you're not in the urban area and your hive is in good shape to begin setting in March. The summer months are a hectic period for bees. It is possible to see them going out in the early morning and returning later in the evening, displaying a variety of pollens. Check the hive, but not often frequently.Swarming is a normal behavior of bees and beekeepers must know the moment to act to prevent from swarming. It is necessary to split your beehive. Beware of hive thieves particularly if nectar is scarce. Certain sub-species have a reputation for this type of behavior. September is the month for collecting your harvest, however be sure to have enough food available for the bees as well as prepare your beehive for winter. It is also the perfect moment to renew your hive.

The dramatic drop in bee populations across the globe has raised questions about the sustainability of agriculture since a wide variety of fruits and crops are heavily dependent on pollination by bees. This is especially true for the more rural areas.An estimated 70 percent of the edible fruits and vegetables are pollinated by honeybees. Pollination is necessary by plants in order to produce, but the vibrant blooms and sweet-smelling nectars make an attractive factor for bees. When a bee gathers pollen and nectar from the flowers of a plant, a portion of the pollen that is collected from the stamens clings to the body hairs. Some of this pollen gets stored on the stigma. This is the the tip of the pistil which is the female reproductive organs that is located on the top of the bloom. This results in the process of fertilization.

In the event that bees disappear from the earth, the whole ecosystem could be destroyed. It is therefore essential to encourage beekeeping, especially in areas where pollination is an absolute need, i.e. those in the rural regions.

The food crops which bees are able to pollinate are apples, almonds beans, berries, eggs, cauliflower, cabbage gourd, mustard, plums, and watermelon peach

The biggest concern for apiaries is the rise of mites. Up until the mid-1980s, certain beekeepers resisted using chemicals inside beehives. But a quarantine breach resulted in the introduction of the varroa aphid, an extremely destructive tick-like honeybee parasite. Around the same time tiny terrors known as the tracheal mites began to ravage beehives across the country particularly in the rural areas.

To protect their bees from other pests and aphids Many beekeepers have turned to chemical pesticides that were effective for a short time. However, the population of varroa varroa-like mites became resistant to the two pesticides that are used to manage their spread, and entomologists found that feeding bees fat patties made from sugar and shortening reduced the tracheal mites down to a level that was acceptable.

Furthermore to that, products that rely on the properties that reduce mites from essential oils (such as Api-Life VAR made of thymol and spearmint and lemongrass Honey B-Healthy) will effectively fight off the mite population in small apiaries. Dusting sugar with powdered sugar is another way to eliminate mites. In conjunction with routine maintenance for hives and breeding bees to eliminate damaged cells, the latest natural methods could reduce chemicals from treatments.

Chapter 4: Honeybee Information

Despite the changing world around honeybees, they have maintained their status for around twenty million years. They are herbivores in their search in search of pollen and nectar however, they can also turn into cannibals in the event that their colony is stressed. Flying around 11400 times per hour is the reason for the sound of buzzing that's often associated with bees. They can fly up to 15 miles an hour. Of all bees, they're the sole one who die when they have stings, with the only exception being the queen.

Every honeybee colony is unique in the scent they employ to enter their beehive. In the event that guards bees fail to detect the scent of their hive, they won't be allowed in. Honeybees possess 170 receptors for odor which enable them to detect their kin as well as communicate with the beehive in addition to locating food. Fruit flies only have the 62 receptors, while mosquitoes have 72, which shows how potent the honeybee's sense of smell actually is. Honeybees don't hibernate like the bumblebees that are closely related. Instead, they are active throughout the winter, ensuring that the nest warm and safe. They make a winter-time cluster, with the queen at the center and then shiver to heat the air. The worker bees alternate to be on the other side of the cluster, to ensure that there isn't a single bee too cold. The middle of the cluster is able to reach 80°F (26.7 degree Celsius) while the outer part of the cluster is between the mid-40s (approx 7 degrees Celsius). This is why they require lots of honey stored for survival.

How are they so vital to the ecology? Based on estimates, about a third of the food we consume by humans each day depends on pollination. Honeybees are the most abundant of any other insect pollinating in the world. Bees pollinate over 90% of the traditional flowers like blueberries, apples, cherries as well as asparagus, kiwis citrus fruits, avocadoes melons, and broccoli.

There are several kinds of honeybees. One of the first steps to beginning a colony of bees is to determine which type you want to utilize. The following list will cover the most popular varieties for those who are new to the hobby, because of their ease of use and their availability locally and through online sellers.

Italian bee (Apis mellifera ligustica) is A well-known bee for beginners This bee is less likely to swarm and also produces plenty of honey. Their appearance is thought to be appealing for beekeepers since they range from light yellow to the lemon yellow. Another benefit for them is that they are less vulnerable to illness than the other honeybees commonly used. On the other hand they can be more active than other honeybees. This could result in lower production if the flow of nectar is not as rapid.

Carniolan bees (Apis mellifera carnica) A popular bee one that is more gentle and less precautions are required. Their numbers tend to rise during spring, which allows them to collect from the first blooms of the season. They also produce fantastic wax combs, which makes them perfect for those who are who are interested in creating candles, cosmetics, soaps and more. They have a strong tendency to swarm however, in order to alleviate crowding, due to the speed with which their population increases.

Caucasian bees (Apis mellifera caucasia) The bee has seen a decline in its popularity over time however, it is still a viable option. The thing that makes these bees distinct is their lengthy tongue that allows them to gather flowers that other species would not be able to reach. They are also thought to be a gentle species of bees. One disadvantage of these bees is their high amount of propolis that they produce and can make their hives extremely sticky, and somewhat difficult to manage.

Alongside those breeds of bees, there's many varieties of popular hybrids that are available. Hybrids are made by combining different lines or varieties of honeybees. The goal of this is to produce bees that are more tolerant and produce more honey.

This list includes the most well-known hybrids in the United States.

Russian hybrids They are immune to tracheal and varroa mites, making them attractive. Because of their inherent resistanceto pests, they can be an excellent alternative to using pest control. They could be a suitable alternative for beekeepers living in areas that experience long, harsh winters since they are prone to rear brood at periods of high pollen and nectar flow. As a result, their numbers tend to change with the changing environment. Contrary to the majority of bees, Russian hybrids will always have many queen cells. This makes them unique in the world of bees.

Buckfast bee developed in the hands of Brother Adam in England during the twenty-first century. The bee was also immune to the tracheal mites. They're also better suitable for the climate cooler of England and are easily located across the United States. On the other hand they are more prone to attack than other species of bees.

Minnesota-Hygienic hybrid - designed to be especially hygienic they are more likely suffer from diseases. Bees with a clean environment are much more likely identify and remove larvae with dead or diseased larvae as well as pupae that have been removed from brood comb. They are also excellent honey producers. They are resistant to American Foul Brood disease, the most destructive and common bee-related disease today.

Chapter 5: How To Setup Your Hives

Do I Need Permission To Keep Bees?
It depends on the place you live. Behaviour is subject to both municipal and state regulations. Cities such as those in the state of New York can prohibit beekeeping, and they may also have by-laws that regulate or ban beekeeping. But, as the trend shifts toward urban beekeeping, a lot of cities are taking down restrictions.

In most rural areas, there will not be any restrictions on the proper keeping of bees. However, certain states require that you declare your hive with the Department of Agriculture. Certain states will check your hives annually for diseases and may also impose limitations on the transportation of honeybees between states, without prior inspection.

If you take the decision to pursue this intriguing activity, be sure to consult your local city or municipality, as well as the state in which you reside to determine which regulations might apply to you. Due to the growing popularity of beekeeping and the growing interest of the public in the environmental realm, it's unlikely that you'll be unable to maintain bees, but it is important to adhere to the requirements of your local area.

Equally important is the need to discuss with neighbors your plans. If your hive's legal and you have a permit, then you're on the right track. But, your neighbors might have concerns - typically because of false notions regarding the dangers bees pose So a pleasant exchange and considerate positioning of your hive can be a great option to minimize any potential problems that may arise in the future. A jar of honey from every harvest can aid in keeping on good terms with people who live around you!

Correctly Locating Your Hives
Selecting the ideal place to set up your hive is crucial because when you make a bad selection, your bees will not prosper, and you might be a problem with your neighbors. The most crucial things to think about:

A sunny area is more beneficial than shade. Bees will be happier and the brighter light helps the hive be easier to study

The hive should be facing east or south as much as you can. Naturally, you do not want your bees to appear straight on the sidewalk or the neighbor's yard, therefore it is important to have these issues thought about too!

The protection against the cold wind is important too Bees must remain warm

It is important to select an area that is dry since dampness can trigger diseases

Place the hive close to your work or home If you can, so that you don't need to go to check on your beehive. Hives that are located near to a home are less likely to be attacked

If you have a beehive on your roof, it is essential to be able to use a sturdy and sturdy ladder to get access to it, as you'll need to transport honey up and supplies down. If it is difficult to access it is more likely to overlook your hive.

If you can you can, keep hives out of fields or crops regularly treated with pesticides

Bees should be within a quarter mile of an water source. If there isn't any water available then you could provide it by putting it in large tanks or bowls with a landing platform made of floating planks. If you do not offer water within the hive, the bees may be looking elsewhere that could be your neighbor's backyard!

The hive must be within about one-mile of principal foraging areas.

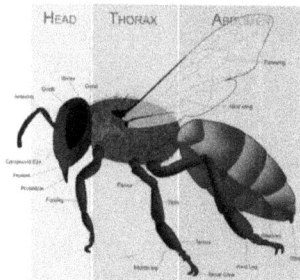

Bees require plenty of blooms for pollen as well as nectar to yield a high honey crop, so a variety of sources are crucial. If there's only a small amount of food

sources in your region, it is recommended to plant some

It is essential that there be sufficient forage available to feed the hive which is why if you're placing your hive in a remote location, make sure that other beekeepers are also using the same forage . There could not be enough forage to be shared!
Pick a location that is peaceful, and away from traffic, children or pets.
The Different Kinds Of Honey Bees
Beekeepers across North America and Europe all have different varieties that are part of The Western Honey Bee (Apis mellifera). There are many regional variations of this bee, which includes that of European Dark Bee (Apis mellifera mellifera) as well as The Italian Bee (Apis mellifera ligustera) and the Carniolan Honey Bee (Apis mellifera cardiac).

Italian Bee image taken by Eran Finkle flickr
The early American beekeepers maintained in their colonies the Dark Bee,

which they imported from Europe with their. For many years, this Italian Bee has been the most popular and widespread bee in American as it is able to produce broods for a prolonged period from the beginning of spring until the late autumn. This means that the number of bees is at its highest in the summer months, and there is more honey produced. This also means you'll have more bees to feed during winter. Italian bees are generally calm and gentle, making them easy to manage. They also are resistant to EFB (European foul brood).

Their biggest drawback is their tendency to roam between colonies. They are also excellent thieves, and can contribute to the spread of illness. The Italians are thought to be excellent housekeepers.

There is a variant that is a variation of an Italian bee, which is known as the Cordovan. The bee is gold in color due to a simple gene, and therefore it is not a new strain, but merely an alteration in color. The benefit that comes with Cordovans can be that they can be easily identified and their color serves as a marker as their replacement by a different queen will be evident when worker bees begin to appear normal with black marks. It also allows you to know the time and place your bees are at work that is a pleasant feeling!

Carniolan Bee image taken by Kam Abbott flickr

The Carniolan is not often kept by beekeepers. However, it is a good choice because it offers the advantage of keeping an enviable population in winter months, which reduces the cost of food, and not having do they rob or strays as it is the Italian does. But, they are likely to be swarming during the spring months which can be a problem. There is a second bee, the Caucasian but it is not often observed in the present moment.

Hybrid Bees

The various regional honey bee varieties of Bees from the Western Honey Bee can breed among themselves. Beekeepers use this method to combine beneficial traits and eliminate problematic ones. The hybrid bees typically generate more honey. They also are more docile, have more resistance to diseases they also have larger broods and might be more gentle. This is known as hybrid vigour.

There's a crucial thing to consider when choosing hybrid bees. We'll see in a moment how your hive is likely to take over the queen's place when she gets old. It is not possible to let this happen when you have hybrids or you'll go through the process of losing the specific characteristics and hybrid vigor that make the hybrid a valuable initially. It is crucial to replicate the hybrid with respect to its parent species every time in order to replicate the unique characteristics of every hybrid.

You must therefore re-queen, which means, replacing the queen by yourself frequently, so that your hybrids are in original form.

Hybrid bees, while more expensive, are an ideal option for those who are just starting out because they usually produce well and are resistant to disease and pests.

The Structure and Functions of a Bee's Body

Bees are insect species, and, like all insects, have three parts. The front part is the head with eyes and antennae that help them find the way.

The middle portion is known as the thorax. It is comprised of three legs and two wings. Legs of the hindswings are equipped with pollen baskets which are used to carry the balls full of pollen from the beehive.

Abdomen is the back partof the body, and has bands of black and yellow on it. Workers have wax glands located on the abdomen. They are which make the wax needed to construct honey combs.

Forgers bringing pollen to hives Images in public domain on wikipedia

Chapter 6: Bookkeeping Of

Beekeeping

Before you begin any venture, it's more important to be aware of the total cost of establishing your business. It is the same when it comes to beekeeping. It's a fact that you can earn a profit through beekeeping, however you can't make money without investing a significant amount of money. A hobby or business cannot be established without a significant initial investment. This chapter will discuss the costs associated with everything you need to set up the first beehive for your newbie. We will also talk about how much honey that your bees will need to produce to be able to break even on your initial expenditure.

If you're into beekeeping solely for pleasure It is recommended to not try to earn a lot of money and instead concentrate on breeding of various varieties of breeds. Earn money by deciding to sell your bees. There is a huge market for bees, and a some kinds of bees are difficult to breed and could cost several hundred dollars for a couple of pounds.

However, if you're looking to keep bees for commercial purposes the focus of your efforts is more on producing. Produce can be increased artificially by feeding bees in springtime, since there is a lack of nectar and pollen in the early spring. The bees will then gather honey and nectar naturally in late spring through summer.

This is a comprehensive listing of the equipment you will need as well as the price it would be to install the equipment. This is merely an approximate estimate and the final price may be higher or lower depending on the equipment you choose.

A complete hive set up Price: $229

The hive is an entire set comprising the bottom board 20 deep frames 2 supers deep 20 honey frames two honey supers, an inner covers, queen exclusion entrance reducer, and an external cover.

Frame feeder - $ 10

Beekeeping equipment- $125

Beekeeping equipment comprises the protective equipment, the netted head as well as the smoking device.

Bees including the queen packagepriced at $110

Books and tools for learning $100

Honey extraction equipment $15

The total cost for the first year for a single beefrom $589 to $589.

Total cost for the initial year of two hives-$138

If you're a novice, and are just beginning, it is suggested to start with two beehives. Two hives can help you to comprehend the tiniest aspects of beekeeping. If you only have one hive on your own, you aren't able to see the bigger picture. It's difficult to differentiate between a great product or a poor one and it is difficult to comprehend why there are differences between two. Two hives can give you greater perspective as they provide you with a choice of what to compare them with!

Chapter 7: Feeding And Nurturing

Bees

This chapter will focus on two basic ideas regarding beekeeping the best way to feed and take care of them. This section explains the type of food you can feed bees, the time you can provide them with these meals, and the best way to ensure they're not being over-fed. The section on nurturing will focus on the actions you can take to making sure that your bees are healthful and secure. Both concepts (feeding as well as nurturing) complement each other since when bees are fed and living in a safe conditions, they can flourish.

If you do not ensure that they are fed well, without taking good care of the environment, as well as other elements of their wellbeing they won't be able to last. Let's find out how to implement both notions.

How to Nurture Bees

To be successful in nurturing bees, it is important to know why bees are essential. Whatever your motivations to become a beekeeper it is important to know the fact that honeybees have been disappearing across the globe. Global warming, overuse of pesticides, as well as habitat loss are a few of the main reasons behind the decline of bees.

There are some who don't understand the consequences of this loss could be tempted to say, "So what?" and don't bother to think about it. But the reality is the decline in the number of bees is devastating for the human race. Do you realize that nearly 1/3 of the entire food crops grown in the USA needs pollination?

The other third is comprised of many different varieties of crops, such as certain kinds of fruits and vegetables as well as nuts. Bees are also the primary pollinators who aid in the growth of fruits and vegetables in the harvest. They contribute 15 billion dollars to the economy that is the United States yearly.

These figures and facts mean you'll be unable to enjoy the fruits of your bees. you'll never be able to appreciate some of your most loved crops that include plants such as blueberries, apples, cherries citrus, cranberries and mangoes. watermelon, plums, pumpkin and more.

So, whatever your motivation for becoming beekeepers (a hobby, for economic reasons or simply because you enjoy it) By fostering bees, you are in is the best way to helping to stabilize nature and assisting in the attainment of a sustainable environment. You can help contribute to the environmental movement by making sure you care for your bees properly and following the actions below will help:

Reduce the use of insecticides.

Pesticides pose a major threat for bees, even though they're supposed to keep the bees free from the threat of bugs. Pesticides can be toxic and include dangerous chemicals that could destroy bees even though you think you're trying to safeguard them. Pesticides can kill all colonies of bees and honey (making it unfit to eat). Some "Biodegradable" pesticides are toxic and, as a result, in lieu of using these many people opt for natural alternatives like praying mantises, and homemade sprays made from pepper, onion or garlic.

Be careful not to be too quick to weed

If you enjoy gardening and gardening, you'll surely be averse to plants, but don't get too quick to eliminate the weeds, particularly if you're keeping bees. Dandelions, for instance, and cloves are adored by bees, and keeping these plants will allow the bees to remain well-nourished. This does not mean that you should remove weeds all. Just remove a few and leave some for bees (you might have new ones to think about you know?

Pick plants that bees will love.

Although bees can pollinate a variety of plants (they are able to frequent the same plant over and over again) There are some plants that they are a fan of. Bees love native wildflowers which thrive in your environment; you can also find flowers from herbs like rosemary, sage, mint and thyme containing lavender, as they are great pollinating options for bees.

Place them near your home and you'll give your bees an incentive! There are flowers like coneflowers, daisies, sunflowers and marigolds, as they are great options for bees. Additionally single flowers are more available to the bees, than double flowers and you might want to keep certain colours of plants. If you've got the same hue of plants it makes it easier for bees to find its preferred.

This is an interesting fact: purple and blue-colored flowers attract more pollinators, so keep this in mind when you plan your garden for bees. Make sure to select a selection of flowers that bloom through the fall and spring, since this will in providing good pollen throughout the seasons.

Find a source of water

Like all species of animals, bees require water for survival. you can aid them by creating a water source that allows them to drink water. The container should be filled with water, and include natural elements like pebbles and twigs (they can sit on them while they drink their water).

Make sure you clean the water regularly so the bees are aware of how to visit every day and drink pure water. Yes, I realize it isn't easy however if you're looking to get the best out of your bees you need to do the effort.

Create a safe and comfortable habitat.

One reason why bee stings occur frequently for some beekeepers is because bees don't feel at home in their new environment. They are agitated, restless and are unable to make high-quality honey. However, your honeybees will be content if they are in an inviting space that is like your home.

This is among the reasons I spoke in detail about the best place to set up your beehive. If you put your beehive in an area that is dry and aren't able to access water, or with no flowers around the bees won't be content. If you have dirty spots, soil and untreated wood in the area to allow them to feel closer to nature, they'll be more content.

Cleanliness and maintenance are essential. Although your bees appear healthy and everything is in order isn't a reason to not wash the honeybee hive. Regular cleaning is essential to ensure the safety and health of your bees. It is essential to check that there aren't any pests present in the colony and the bees are in good health. I will address the issue of controlling disease and pests in Chapter Seven.

However, prior to getting to the next chapter, you must have an annual maintenance plan for your hive that is based on advice from beekeepers who are experts and knowledgeable. Don't wait until it's easy to clean and maintain your hive prior to doing it.

Learn to assist an exhausted bee

It is possible to nurture your beehive through constant keeping an eye on the bees. And If you notice an insect trying to fly be aware that it may be wounded or is about to die. In some instances it could be that the bee is just be exhausted. In order to help it regain power mix 2 tablespoons of sugar with one teaspoon of liquid. Mix the mixture onto an ice cube or on a spoon and invite the bee to consume it.

The mix should not contain honey or sweeteners that you have on the kitchen cupboard or pantry, and you should place the mix in a spot near by so that the bees have access to. If you spot bees resting on flower, it could be weak. It could be in a resting position on a flower, so keep an watch. Keep in mind that these creatures are not human that cannot inform you when they're sick, but by continuous monitoring and examination it is possible to tell that something is not right.

Make yourself a champion of bees.

In your own manner, it is possible to support bees' cause by informing everyone about the consequences of declining bee populations on our environment and economy. Bee-related organizations across the country have members that are adept at promoting bees due to various reasons. If you are a member of such groups you are able to lend the voice of your community and help contribute to your contribution to keeping bees healthy.

Keep in mind that if nobody does anything, and you and I aren't then you won't have honeybees to feed. If you're a member of the beekeepers' organization within your area, make use of their platform to help people realize the importance of this. This isn't an attempt to convert you into become a "Bee activist," but it is certainly a worthy cause that you ought to consider. Talk to relatives or friends. Some may not know why you're keen on bees. So let them know by sharing what you've learned about the effect of bees on nature as well as the economy.

Feeding Bees

While allowing bees to collect their food is good however, it is important to know that you can make food for them in the event that, because of reasons other than yours there aren't enough sources of honey in the world. It is possible to make their food source if they're not able to create it and especially during the first week they are in the beehive. There are many types of feeds available, including sugar mix and feeding times.

Step one: Select one of the feeders.

There are many kinds of feeders. Some of them are:

The frame feeder

It is possible to use frames, which are conventional feeder made from plastic or wood. The feeder you choose should be within the 1-2-gallon range and have a sliding feature that allows bees to slide up and over the sides. Additionally, you should get an air-tight feeder that has floats in order to stop the bees from drowning when water is poured in.

These models are available at the most popular bee shops and ensure you use feeders which are simple to operate (you can easily open them to change the feed and clean them quickly too).

The Boardman feeder

To avoid allowing your bees to drown, choose frame or miller feeders for wooden feeders that resemble big boxes. They can accommodate mason jars with food upside down. Just put it in the entryway to let the bees enter the pot to drink the syrup. They are also simple to handle. To fill them, you'll need shake them gently.

The feeder that is inverted

The feeder functions as an water cooler, which functions as mason jars that are inverted, so you can put it in the top of the entrance, so the bees are able to fly through it. It is possible to cover the door with a cloth to ensure that the bees do not hover in the jar whenever you need to fill it up. By using such a feeder, the honeybees will have a lower chance of drowning.

The miller feeder

If you're looking to manage a large number of honeybees this feeder is going to fit the bill as it's a plywood feeder that is larger than the standard feeders. Miller feeders allow the bees access to the feeder which allows them to move across the hive using the feeder. With this feeder, you'll be able to store more syrup. Make sure you clean the feeders, change out any spoilt food items, and keep the bees in good health.

Step 2: Get ready for entry into the feeder. For entry into the feeder, purchase cotton suits to ensure convenience, and give you the ability to move about and keep bees out. Find a suit with knee pads, leg zippers and elastic wrists, a gusseted crotch and a double-ended zipper. The suit should also include a an area to store tools and of course, it should be sting-proof. The cost of these suits differ based on the design and quality. Therefore, you should only purchase what you can afford.

Then, ignite the smoker with a lighter from a cigarette or match. However, do not cause it to get too hot. Inhale a few puffs of smoke towards the entry point and on the uppermost part of the feeder because this can disarm the bees by releasing hormones. After waiting for a moment after each puff before opening the hive. And If you observe that the bees remain in a state of agitation, you can take another puff. Be sure to follow the directions in the on the manual for the smoker you choose to use.

Step three Step three: Mix the syrup

The syrup is a good option when you don't have honey available to provide the bees with food. It's easy to make as it only requires water and sugar.

*1 pound of sugar, or five pounds sugar

1. 1 Quart of Water, or five quarts

Add the sugar to the hot water (use the measurements according to how much or how much you wish to create). Then, you must take the water off the heat prior to adding the sugar. Then, after this, let it to cool before serving the bees.

Step four: Plan the feeding times.

Finally, you must create a feeding schedule. The winter months are usually the most difficult time for bees as they aren't able to get food, and they become hungry. It is important to create an eating schedule that will allow you to feed your bees through winter. Create the recipes for feeding syrup in September or August (you could also prepare this in the month of October but don't do it till December).

You can ask local beekeepers for advice if you aren't sure of the amount of food you need to keep in the freezer for winter. You can then use the seasonal mixed syrups to make gallons of food. Give the hive meals when they require it, so that you don't feed them too much.

Use your calendar to determine how you'll feed them on a daily basis using the information you collect from other beekeepers who have experience, as well as your observation of the bees' eating habits.

The right way to nurture bees with a sustainable end with a sustainable goal in mind will allow you to not just increase the number of your bees but also increase the number of bees you have. They also aid in the global efforts to conserve. You don't need to be under pressure for just talking about conservation. I would like you to know that your efforts are important, and you're doing more than just cultivating bees.

If you are able to provide food for your bees and care for your bees, the better you get proficient at it, and it becomes regular part of your life. The initial few weeks of keeping bees can be challenging particularly if you're on an active schedule However, don't fret about it that as time passes you'll get used to it. We'll now learn more about the bee season at Chapter Five!

Chapter 8: Common Honeybee Species

North America has exactly zero honey bee species native to the region. Honey bees were introduced into the new world by the first European colonists in the 1600s. The first kind of bee which was introduced to the continent is that of the German honeybee. Since then, a variety of varieties have been brought in. There are currently six honey bee species that you might encounter during your journey to beekeeping.

The species you choose to use will be contingent on a variety of aspects. The average temperature in your area as well as humidity, elevation etc.. All of these factors contribute to which honey bee will be most suitable for you.

Each species has advantages and disadvantages, which must be considered in your specific situation. Certain species are better equipped to fight sickness, while other species could produce honey that is more appealing, and so on..

The following summary of the most sought-after honey bee species will help to understand the options available and provide you a framework for conducting more thorough study. Do not be afraid to seek out people who have more experience and get their opinion.

The Buckfast Bee

Buckfast bee Buckfast bee was created through the efforts of the English monk during the 20th century to supply English beekeepers with a specific species of honeybee that could cope with the extreme levels of humidity and moisture that are typical of the island. The extreme weather conditions made bees with a greater likelihood of contracting a disease, and less capable of defending themselves from the spores and other issues.

These bees are sturdy and are able to withstand the most challenging environment. They're slow in their production of honey and are thought to be somewhat defensive.

The Carniolan Honey Bee

Carniolan honeybees are famous as gentle and easy to be around. Many beekeepers discover that they don't need for any type of protective gear while working with this species.

Because of this docility , and it being the case that bees is a master at making honey, They are one of the most sought-after bee species by American beekeepers. Carniolan bees tend to reproduce rapidly during the spring. This is why they are so effective in making honey. They are also more likely to gather towards the end of spring in order to prevent overpopulation in the honeybee hive.

The Caucasian Honey Bee

The once popular choice of beekeepers The Caucasian bee has been pushed out of fashion in recent years. Caucasian bees are excellent in making beeswax, and are known to produce more of this ingredient than other species.

These bees are also renowned for their gentle nature, but they don't have the capacity to rapidly populate the hive when it is in spring. This usually results in lower honey production and a less likelihood of having a swarm.

The German Honey Bee

Sometimes, referred to as"the "Black Bee," German honeybees were among the first to produce honey bees in the new term. They can be very easily irritable and have a been criticized for being difficult to deal with. German honeybees are believed to be robust and can withstand temperatures that are colder than normal however they are more vulnerable to diseases as compared to other types of. They were once the most preferred breed of bee to raise but it has gone out of fashion in recent years and is today thought to be very rare.

The Italian Honey Bee

It is believed that the Italian Honeybee is the number preferred choice for beekeepers and for good reason. The species that was brought to in the United States in 1859 is well-known for its ability to maintain a large population as well as being relatively easy-going and resistant to disease and pests. They are also extremely efficient in producing honey. One of the peculiarities that comes with Italian honeybees is the fact that they can hear the cluster's sound echo across the walls of the hive in the winter months, while other varieties are more quiet. This can be particularly beneficial for a newbie beekeeper who is wondering if their honeybee hive can withstand the cold.

It's not just about skittles and booze when it comes to the Italian honeybee. However they are known to use more honey than others (leaving less to the bees) and are also known for taking other hives' honey their honey. A lot of bees steal honey from other hives to ensure that they have enough to last through the winter months. The honey they steal increases their exposure to pests, parasites and diseases.

The Russian Honeybee

Russian Honeybees possess a distinctive characteristic that sets them apart from other species we have examined. This kind of bee is extremely effective in fighting common mite spores, making it a popular choice for many beekeepers. They are known for maintaining an hive that is cleaner than other species of bees. That is among the reasons they are so immune to diseases and infestations.

This is among the most recent varieties of honey bees that are that is being developed in America. They were quarantined for quarantine by authorities from the United States Department of Agriculture until 2000. They are now open to beekeepers who are hobbyist or commercial.

Overall honeybees from Russia are moderate producers. They display a range of characteristics that make them unique and suited for a wide range of habitats

Make sure you consult an experienced beekeeper who can determine what kind of honey bee is best for you and your requirements most. The choice of a species of honey bee that you want to raise is a decision that needs to consider a myriad of variables. It's a crucial decision and shouldn't be taken by those who aren't experienced.

If you are spending longer working with and studying bees, you'll surely be inspired to take a shot in keeping different species of bees or even using various kinds of hives. This is what makes beekeeping enjoyable and instructive. Do not try to tackle all at once! Be patient and enjoy this process. it will not "bee" long until you're an expert and successful beekeeper!

Chapter 9: Care Of Your Bees

If you want to be successful in keeping bees, be sure to do the correct things at the appropriate time of the year. Honeybees are in essence wild animals, and they react to the changing environment and seasons. It's our job to keep pace with our beesand and not the reverse. To improve your health and the well-being of your bees as well as your honey production, you must follow these guidelines:

Food Stores

In all likelihood your bees will be able to sustain them through winter months, but as they approach the end of the winter season, they're most likely to be low. If the winter continues longer than expected your bee colony may die if it can't find food. So, it is important to check your beehives to ensure they have enough food. You can check this simply by lifting the beehive off one side and weighing the weight. If it's too light, you can supplement the bees with sugar or fondant syrup. Make sure to increase the colony's capacity to make it through the beginning stages of colony expansion. When you reach the season of spring, they'll be foraging for themselves.

Building up the Colony

It's not something you have to be aware of however, you must be aware of. The queen bee stops making eggs during winter months, mostly in January and December. However when the days start to get longer then she begins again and allows the colony to create the new brood. As food becomes more easily available to the bees and more eggs are laid. And, while you could have begun with 5000 bees by close of winter when the middle of summer comes around it could be anywhere between 40 to 50,000 bees within your colony

Inspecting the Hives

In the springtime, as the weather starts to warm it is time to begin the hive inspections. It is best to do this when temperatures rise to the temperature of

15° Celsius or more and there isn't any wind. There's an old saying "if it is too cold to work in rolled up shirt sleeves, it's too cold for the bees". In the event that the sun's bright and there's no wind, it is possible to look for bees during colder months, but for only a short time. If you're in any doubt, do not take the plunge since a chill could destroy the brood.

These inspections are a good way to assess how healthy your colonies are. Check to see if your queen is pregnant eggs - you should see an egg pattern - and also, do the larvae of bees look attractive and healthy? They should be white with a pearly appearance and shaped like an "C." Is there enough food sources for the colony to survive? You can identify the queen during this time in case she isn't completed.

If there aren't eggs, then there's likely that the queen has left. You'll either need to find a replacement or merge the colony to one with queens. If you have eggs but they're not being laid regularly, it could be a problem with the queen. In this instance you'll require the replacement queen.

The next inspections will be mostly to keep the condition of your colony, monitoring the brood count and also warming management.

Spring Food

In the event that your honeybees access to crops such as borage and rapeseed you might be blessed with huge quantities of honey. the spring harvests which are the most ideal season for the production of honey. If you're not close to these kinds of crops it is important to plan accordingly . Find farms that grow them and inquire with farmers if you could put up your hives for a period of time. If your bees live close to these crops, they should be prepared for a harvest that is early. Rapeseed honey must be extracted prior to the time that the bees enclose it, so that you can get it immediately.

Controlling Swarms

Bees are able to swarm at any time from April to July. It's an natural method used by bees to boost the number of their colonies. If your colonies are allowed to overflow they will start to raise queens from the eggs of their hatchlings. If a queen is born and is placed in her cell, the current queen will leave with her, possibly taking as much as half your colony. This may keep new queens from forming and the new queen will rise to become the leader for the entire colony. But, certain colonies exhibit strong swarming characteristics which means that you may be able to have a number of new queens. The majority, if not one which will go away from the colony, bringing many bees along with them. This can leave your colonies diminished.

Swarming can be beneficial to beekeepers, since it lets them divide their colonies and start new colonies, it could be trouble when a swarm is lost and results in a dramatic decrease the production of honey.

Re-Queening your Colony

The queen bee's life span can be up to six years, but it's not often. Queens who are older don't do as well and are typically replaced by a new queen. Beekeepers anticipate a queen to lay lots of eggs in two seasons which is why in the second one you need to plan to introduce an additional queen. It is recommended to do this when the summer is over and the autumn season begins which gives the colony the chance to create enough bees to be able to reproduce before winter arrives. Some beekeepers decide for spring in order to make this decision, however.

A method to re-queen an existing colony would be to collect frames with an ant-queen cell and place it in the nucleus box along with several frames of broods sealed that come from at minimum one hive. The nucleus will then be complete with a comb that includes a brand new foundation it. It is possible to monitor the queen's new status to observe how often and efficiently she eggs and her personality is similar to. After the season is over the queen could be replaced by the previous one.

Honey in the Summer

After the flowers and crops of spring are over, there is likely to be a decrease in food to bees. In late June, summer crops start to bloom. There are many people who will not be able to harvest a decent honey crop in the summer It is contingent on the available forage to the honeybees. When you've plenty of flowers in your garden and clover, you'll be able to harvest a decent amount. If you do end up with an impressive harvest, be sure to remove the honey before August. In the following months, the bees will start to build the winter food sources.

Heather Honey

This is a honey crop which can be harvested over a relatively short time during the summer's end. If the weather is good colonies can produce substantial amounts of sweet honey by harvesting the flowers. However, as with the spring crops planning ahead is essential to locate the best locations and get permission to set up your beehives. Honey from the heather is generally valued more in terms dollars than other kinds of honey.

Winter Feeding

As summer draws to a close it is essential to make sure that your bees are stocked up with enough to make it through winter. If you're skilled enough, you are able to take the hive off to one side to determine the weight. As time passes, you'll be able to gauge the reserves of the hive based on its weight. In the ideal scenario the bee colony would be able to fill the frames to the frames that are outside with pollen and honey, and the ideal weight is approximately 40 pounds.

During the season of active The brood body is utilized by bees to make new broods and keep the honey that is not used up. The beekeeper is responsible for removing any supers (containing the honey) as well as the bulk of the stored surplus. If carried out at the right moment, the colonies can build new stores to store for winter. If you do not finish it late or the weather starts to change early it is necessary to lend the bees some help.

Utilize sugar syrup in feeders to enhance the food. It is made up of equal quantities of water and white sugar (brown sugar is contaminated with toxins it that could kill bees). If you aren't able to feed until the middle of September the syrup must be more concentrated because the bees are able to make it more quickly. Make a decision based on whether the bees require help or not and continue feeding until your hive is large enough.

Disease and Pest Control

Bees are susceptible to certain bugs and diseases that must be managed:

Varroa Mite

They will attack larvae as well as adult bees in a bloody attack, sucking their blood away and weakening their defenses. If your bees' wings or legs missing, it is likely there is a varroa problem in the beehive. Utilize disinfecting medications in the later the summer and into early autumn. avoid using when harvesting is in the near future or when honey flows. Keep levels of mites down as bees enter winter. Otherwise, you might not have colonies by spring.

Foulbrood

The cause is Bacillus larvae, and is able to kill significant quantities of broods that are sealed. European foulbroods aren't quite as severe, however it can nonetheless cause larvae to deform. Disinfecting medicines is best sprayed between spring and late in the autumn and not during honey production is at full swing. If you are unable to stop the colonies from dying, you'll have no choice but to kill it and start new.

CCD

It's unclear exactly the mechanism by which CCD (Colony Collapse Disorder) is brought on, but it's believed to be caused by a fungus known as Nosema Ceranea as well as IIV (invertebrate Iridescent virus) within the hive. There is only one solution, which is to eradicate the colony and then burn it along with the honeybee.

Maintenance

Winter is the perfect time to conduct the maintenance of hives on spare hives as well as components of the hives. It is possible to cover your hives when you experience unusually cold winters.

However, it isn't recommended since it can keep your bees in. Make use of an entrance reducer assist the bees to defend their hive from rodents and other predators, but ensure the opening is 3/4 inch and measures between 3 and 4 inches in length. This can also prevent larger colonies from becoming dominant in spring.

Chapter 10: Preparing For The

Honey Harvest

The typical improvement in a honeybee settlement in the time of season was described before. We will now look at the most important mediations that the beekeeper can make during this development. The purpose behind these is to ensure that the settlement gets to the highest quality, at the moment of the most pristine honey stream. The measures discussed here are only applicable to hives that have free casings. It's hard to interfere with fixed brush hives.

Thorough Inspection

At the end in the dry season (in the tropics with sticky soils) or during the dry time (in the dry tropical zones) the actions of the colony increases. The new season of honey bees is about to begin. It is time to begin the new season by assessing the settlement of honey bees. First, you must determine the presence of queens still present, and if she has produced any drone brood in the specialist cells (swelling cells capping). Queens that are slanted to reproduce is a rambler. (either due to age or lack of matings) is evicted and killed. The colony with no queen is joined to one with queen. If only a few honey bees that were part of the initial settlement remain, it's best not to join them with a different state. In all likelihood honey bees may be suffering and, if they are removed, they must be killed and eaten. If the honey bees' hive's floor is protected by wax scraps or precious sugar stones, they should be cleared away (consider the wax moths as well as Ants).

Cleaning the baseboard clean isn't a problem. For hives that have an attached baseboard that is not removable, it is best to put the casings in a clean chamber. Since the size of the state is typically tiny, there should be no difficulty in finding the queen. The queen's wings must also be eliminated (why this must take place will later be explained in the section regarding the counteraction to the swarming). Make sure your hands are clean. Make sure you hold the queen from the back of the neck (never in the midriff!) and then carefully cut off some of the wings on the front. It is recommended that you first practice this with a few drones. To make sure you don't cut off unfertilised queens (which could be making the mating flight) this part could be cut if there's an over-sized brood of a specialist! After you put the queen back onto the casing from which you found her, make sure she's not disturbed by any other honey bees due to the scent she acquired from being taken care of. In the event that it becomes essential, put a few drops of smoke. In the event that honey bees are extremely irritable and irritable, you should place them in a piece of a queen pen that is closed with a ring of sugar batter (sugar combined with honey) and place it in between casings. The workers will devour the treats and thus free the queen. If you are unable to locate queens to confine the queen, you should rely on an empty match-enclose that you've made some gaps. The honey bees , in turn, will let her out of this.

Expanding an existing colony

There are various ways to increase the size of the size of a settlement. Consider that one massive colony can produce more honey than smaller arrangements (it will be more effective financially). It is recommended that you, in this way, try in the event you are aiming to collect as the amount of honey that could be expected to concentrate on the most important possible settlements, not necessarily the most effective number of states.

A bee may be increased through:

The growth of the brood's home size This could be done by enticing when conditions aren't ideal and by trading outline (between both the super brood and the brood chamber) and by giving additional edges.

Joining states In the early part of each period, you can join small settlements successfully by sprinklering sugar water over both colonies at night , and then placing the casings of the two provinces on the other hand , in a large brood chamber. The most experienced or evidently awful queen is removed. If it isn't important to you, which queen is left then join the US. On the next day, there will only be one queen left, and mostly the one with the most youthful. Inconspicuous casings that contain food are stored away to ensure that they are safe from honey bees and ants (in an enclosed brood chamber or box, paper or plastic bag).

Wirework It is possible to use an air-ventilation screen to allow two settlements to be acclimated to each other in a controlled manner. The casing is made of wood housing, with identical dimensions outside that the inside spread. The edges have fine wirework running across the underneath. One of the four bars, you can make a mark 5cm wide and 0.7 centimeters deep. This opening will serve the opening for the flight entrance.

If you are able to shut the hole in the screen for ventilation and pivot it, you can use a piece of wood of the same size to the size of the indent in the bar. When the screen is empty this tiny entranceway fills with flight load. In the case of hives in which it is the case that honey chambers are situated next to the brood chamber (and not above it) create a casing that is comparable to keep the two separate. In this scenario it's not required to create a flight entry at the edges, since the honey chamber is equipped with its own entrance to flight. The ventilation screen should be placed over the hive that is part of the main colony. Then, place the hive of the state to be joined on top of the voyaging outline. After seven days, you are able to remove the screen and go into your brood home. This will ensure that you have a more ball-shaped brood-nest.

The amount of honey produced by a colonies can be increased with:

By adding outlines using topped brood from various provinces (after you've passed or tapped honeybees). Be careful not to use too many casings of the family that is topped and the settlement should be able to share all the edges of the family to ensure it is warm. The family is likely to die.

The honey bees from a neighboring colony towards your hive to be strengthened. The hives to be joined may have been in close proximity for a few weeks. After that, you can you must move the hive from the state where you want the honey bees flying to fly 5 meters from through the colony's hive.

You must build the most of your way to the state of the main honeybee hive. The honeybees that fly in the first hive currently be able to join the hive next to them, which is in the event they're carrying honey, or even honey. In the event that they do not, they are restricted by the gatekeepers of the colony that follows. This way, let honey bees flow through during the wonderful flow of honey. The settlement that was refused its flight honey bees needs to be fed for a time, if the storage of honey is overly small.

Migratory beekeeping

The honey bee hunt plants that can yield an adequate amount of honey are usually distributed across the countryside. In the event that they all bloom simultaneously, they won't be able to benefit from the entire range unless you've spread your hives across the entire region. In the event that these honey bee hunter plants continue to bloom and in any event you will be able to collect more honey through the honeybee state. They must be encouraged to the point, however it is, you could choose to go with honeybees with wooden casings. Choose a location that is, or could be protected from insanity, burglary steers, game, and fire. Also, you must make sure that the honey bee scavenge plant aren't sprayed with harmful synthetics to honey bees.

The preparation of the colonies to the journey

Prior to the journey The chambers and casings shouldn't be loosened. The honey bees have been shackled with the various parts of the hive which is a desirable location for transport. It is also possible to take another brood chamber, or honey super. The day prior to leaving replace the crown load with the ventilator screen. Secure a couple of strong ties or ropes around the hives to ensure that the baseboard or weights don't be moved during the journey. They are best made by plaited jute or sisal. You can also create them using 10cm-wide pieces of the inside cylinders that are worn out in a tractor or vehicle. The flight entrance should be closed at night or in the early morning when it's not but dim and honey bees aren't flying. You could use a small wooden board or (even more effective) wirework you attach to the top to the entryway. A piece of froth elastic that is inserted into the flight path is also a great idea all-around. If there are honey bees that are holding the flight load and you want to get rid of them, add water mixed in with a few drops of vinegar or smoke them to force them into the beehive. When transporting the colonies, they should be placed in a way that the casings extend in a parallel fashion to each other along with the width of the car. This will ensure that the casings won't crash against one another when there are sudden shifts in speed. The hives need to be permanently attached to the car. It is essential to accelerate and brake slowly and drive with a greater speed. Include a honeybee cap smoker, sledge, pincers, and some nails, similar to some water. There is a possibility that the settlements can become overheated in transport, and you can stay away from this by showering them with water. It's ideal for transporting honey bees through the most pleasant time during the working day. It is also possible to hang jute sacks of wet jute over the hives, so that the colony gets cooled but the air supply isn't interrupted. Bring something (earth or the froth elastic) to close all the gaps might be discovered in the course of your excursion.

As soon as you arrive at the location you will need to place the hives on a rooftop or tree to protect them from the sun and rain. When you are done, put the crown board on top of the screen of ventilation.

The honey bees will then walk towards the downwards. Once honey bees stop it is possible to access the entrance to flight. In the future you will be able to remove the outline for voyaging and replace that with the crown board. Drink water when essential.

Swarming

Swarm cells work during times when brood homes grow. Honey bees are less active, and fewer honey bees venture out to look for food for food, and no new brush is created. The state is set to segregate itself , and could do so at least a couple of times, but only the time you sit in meditation. A good beekeeper is aware of the progress of the nation and tries to stop swarms at a given time, despite the reality that this will never be successful.

Take note that you don't have to stop swarms, especially in areas where it's not difficult to stop large swarms. If you join these at the appropriate time, you'll be able to gain stable states that will assure a high honey harvest.

States that have been swarming are slow to collect honey. These settlements are not much significance for the season's honey gather.

Methods to defer the swarming.

Provide ample space within your brood room. One of the primary reasons of swarming is that it is a lack of space. The additional space can be provided (draping empty brushes inside the brood chamber, or making a brood chamber) The brood house can be expanded which keeps the queen and members active.

Eliminate topped brood. Take a few casings out with top-of-the-lined family, then tap the honey bees away from the edges of the brood chamber, then put a few tips in balance using empty brushes. The sides that have been removed with brood may be offered to states with fragile constitutions. In this way you restrict the growth of the country , and at the same time allow more space in the brood's house.

Let honey bees fly into (see Partially joining) The idea is good method of reducing the severity of a condition that is slanted and swarms, as well as to expand another colony by introducing flying honey bees.

Clap the queen. In the event that you cut one of queen's wings towards the beginning of the period when the colony is small, she will when she first attempts to depart with a part or the entire state land on the ground and chew the pollen. If the swarm realizes that there is no queen it'll go back to the colony. Since the queen that was previously in place is possible to be identified by the hive once the queen cells of the first queen have been fixed and fixed, there will be at minimum one queen within the colony in seven days. This means that at the very least, one of the auxiliary Swarm will begin to take off. If you've seen the queen trying to swarm, you could cut off all the queen cells, but not until next day. To ensure that you do not let more queen cells go, examine all edges carefully, before removing or tapping honey bees. First, you must check the side that is adjacent to the queen cell you want to keep, and then deal with it in a cautious manner. If you are drawing honey bees out from the casing, it might harm the young queen inside her cell.

Methods to avoid the spread of swarms

This should be done once you have enough drones, but only after there are the swarm cells. If you follow the appropriate steps you'll be able to control the amount and dimensions of the swarms. If you do this your chances of ensuring that you'll lose one or all in the settlement are dramatically reduced. In the event that you create an artificial swarm about a month and a half prior to the major honey stream then you will then be able to establish a colony that includes many flying honey bees during the time of the stream. Additionally, you'll have a small settlement (with the queen of old) that is available for the future. There are several methods to make fake swarms. But, the basic method approach is equivalent that the colony is divided into twoparts, one with the queen, another with the artificial crowd and the other without, but equipped with the an actual queen cell(s) as well as brood brush that contain eggs and young hatchlings from which honey bees are able to create fresh queen cells.

In separating colonies is crucial to plan the process carefully and consider the following focuses which include: what are the conditions required for the new colonies and the best location for them, which one is the previous location that is where all honey bees flying return and which one in each other location that only the young honey bees remain.

In general there are two options:

The queen of old is still in the old location, and this is the area where all the flying honey bees will reside. To deal with this the beekeeper should remove all brood brushes that are open and also the top brood brushes except for two of them, and move them in another hive just a few meters away. The new colony honey bees will create queen cells from cells with recently incubated hatchlings. There is no need for empty brushes for this colony, neither are sheets for establishment, since there's no queen to lay eggs or honey bees that fly to gather pollen or honey. A couple of casings with a baby, and an edge containing some powder are needed to sustain the hive until the queen has laid eggs, and honey bees are forming into honey bees that fly. In the previous settlement, the queen would keep laid eggs, and no brushes and sheets for establishment are needed. Honey bees that are young emerge from the brood with the top that remains, and will provide the hatchlings with food. While they are at it, the flight honey bees continue to bring honey and pollen.

The queen of the past is relocated to a new location. In the new hive, she will remain with the young honey bees which need to be shaken off brood brushes. The most efficient method for doing this is to place the casing with one hand over the hive that has just been created and after that, tap it in a slack manner with the other hand until the honey bees fall off. 66 percent of the brushes need to be handled this way. Keep in mind that a substantial part of the honey flying bees are likely to return to the hive they came from and consequently, the new colony needs to be filled. Place the establishment casings as well as void brush outlines inside the new hive along with honey and pollen frameworks to allow the queen to keep from creating eggs. Avoid placing brood brushes inside the colony. They must be kept in the province that was previously used where the state will construct new queen cells.

Utilizing these two methods with the two methods, some queen cells will be released in the hive but without queen. In order to eliminate the possibility of swarming it is recommended to check the settlements every 9 to 10 days following isolating , and then to eliminate all queen cells, with only the most powerful one, which is where the queen will emerge. Queen cells which were removed are able to be used in various colonies to replace the queens that have died. Make sure you handle an emerging queen cell and ensure it is in good standing. With a pin that is required or needle the battery may connect to the colony's brush that is not ruled by having a queen. Make sure that the territory in which you place the queen cell is not ruled by a queen at least one day, but not more than three days. In the correct manner, this will make the state recognize the battery's new status.

If you have to separate the state, and you are unable to locate the queen, it's possible to divide the colony into two identical sections. The first colony is located in another location with a growing number of honey bees (intentionally removed from beekeepers) and the other in the old location with all-flight honey bees. After 4 days of observation, beekeepers is able to look over the two colonies and find out which areas the queen is in (look to find eggs!) and then re-arrange both colonies according to the highest standards. When re-masterminding, make sure that you don't cause the queen to move unintentionally. The final option is a sensible arrangement when queens are difficult to locate in large, guarded communities.

Absconding

African honeybees are known to quit their hives leaving rather than by the normal swarming procedure. They disappear due to the fact that they don't have enough food or water, or due to perturbing influences created by beekeepers. They are, for instance very sensitive to the brood's outline which are being evacuated. In addition, others, or other animals may trigger sudden departures. Honey bees generally try to find the best place to conquer their obstacles. This is also an advantage for beekeepers. The states move generally in huge numbers every year in synchronized techniques, beginning at one area and moving on to the next. The beekeeper can try to determine where, how and when the typical of honey bees disappearing is changing. By placing bee sanctuaries in the way of the disappearing out honeybee colonies the beekeeper can be assured that they will receive the same amount of settlements that vary. The best way to stop colonies from the possibility of slipping away is to safeguard honey bees from unsettling elements and ensure the bees receive a certain amount of nutrition (in all cases, at least four or five brushes stuffed by honey) is provided.

Chapter 11: The Bee Harvests

The most appealing aspect of the relationship between a beekeeper and bees is the rewards of collecting time. The majority of the things that produced by honeybees is of value to the beekeeper, including honey dust, beeswax and much more. There is no greater satisfaction in the taste of that first honey you harvest from your bees. It's a taste which is unique for your specific region. In this segment you'll learn ways to use the products from the hive to make other enjoyable products, like candles and skin-healthy basics. There are also different methods to use items you have in your hives, all created by the bees that have successfully utilized in your commercial.

Results of the Hive

Many people believe that honey is the primary product of the honeybee. But, a hive can produce numerous amazing products that can be used in amazing and varied uses. Honeybees convert their collected assets into products that people can take, eat, or make use of. It's a wonderful connection if you're careful to leave enough to meet the needs of the state when you gather these items.

Honey

Beekeeping is the idea keeping bees in order to collect honey. Honey tastes delicious and beekeepers are generally grateful for the honey they receive from their neighbors however it's intended for personal use or in the unlikely event they decide to market it. The United States, the USDA offers 3 honey assessment forms (A B, A and C) which are used for determining scores for different aspects:

1.) Moisture content amount of water

2.) Deformities are not present The absence of propolis, particles and dregs

3) Flavor and fragrance Taste and scent: derived from the primary flower source

4.) Clarity: simplicity and no air bubbles

Honey is also a shaded product that do not affect the evaluation, however, they discern the honey's taste by comparing light honey to soft and darker honey being more grounded:

1.) Water white

2.) Extra white

3) White

4) Extra light golden

Five) Light golden

6) Amber

7.) Dark golden

Most honeys contain peroxide activity that is one of the reasons of the reason honey is antibacterial and won't harm if stored dry and in the same container. Honey is also hygroscopic, that means it can without much difficulty hold in dampness the air surrounding it, which makes it crucial to remove dampness from it.

Beeswax

While bees make use of beeswax for storing honey inside cells and protect pupae and hatchlings in the process of development and pupae, you can use beeswax to create a variety of special items, like candles and skin care products that are healthy in addition to being an additional ingredient for food items. Regal jam is delivered to the hypopharyngeal organs of the medical attendant bees. They provided care to two young hatchlings as well as the mature sovereign bee. Bees don't keep the illustrious jam. It is constantly kept in mind for fresh. When raising sovereigns, the bees are able to transfer an excess amount of the eggs to hatchlings from sovereigns and whatever remains un-eaten will be gathered at the bottom of the cells. Because the sovereign has larger Ovaries and is living longer than drones working and drones, there was a lot of speculation in the 1950s suggesting that people could consume imperial jam, they would become more mature and appear younger and live longer. However they were not proven.

Propolis

Propolis, also known as bee stick, is made by honeybees, who collect the sap along with other pitches of trees or plants, and mix it with beeswax and spit. It is used to fill in gaps and holes within the honeybee hive. Propolis can be collected from a hive by rubbing it off the edges or dividers, or make use of a propolis-screening plastic that is placed inside the beehive. Numerous stories describe the various medical benefits of propolis such as helping in treating sore throats, colds pimples, cuts, ulcers as well as consuming substances, and other. A company also makes toothpaste using propolis to nourish and strengthen gums. Dust is packed with proteins, nutrients and minerals. Many people have reported that taking small quantities of dust in the vicinity daily causes them to create personal vulnerability to sensitivities. A few dust-related products claim you've demonstrated increase in white and red platelets, a decrease in cholesterol as well as reduced the triglycerides.

Bee Bread

Bee bread is a protein-rich food and is a popular food for people. It's a crucial source for well-being and endurance during winter. Bees that search for dust collect it and then return it to the beehive. They dump the dust easily into the open cells near the brood as well as close to honey stores, creating dust bands and are often referred to by the name of bee bread. Many new beekeepers discover cutting brush to be an amazing way to start to pursue a new interest. It's not as tense or as difficult as the process of separating wax or honey, and cutting brush has many health benefits when consumed. If you want to market items the honey you collect the honey from your hives. Starting with cutting a brush is the best option. If you cut the Dust brush honey, it is recommended to put it in a freezer for at most at least 48 hours. This kills eggs and prevents them from laying eggs within the bundling.

It also kills the moth eggs that are in the drawn brush. However it doesn't stop moths from gaining access to your brush to lay eggs again more. If your space for relaxation is a good fit, you can keep the brush that you cut inside. There will be some limitations in the event you attempt fixing the brush into plastic bags and then putting it in your cooler. Province and you tend to bees and are able to split in a new hive then you could sell the next hive as well as keep it in order to provide more frequently honey and various other products. Cash is made through the sale of bees, by splitting an hive or selling sovereigns in nukes, bundles, or larger hives are difficult to achieve a successful effort for a beekeeper who has an expanding apiary.

Setting Up Your Honey Extraction Space

The process of extracting honey is one that that your entire family will enjoy. With just a few items, you can bottle the honey you extract your own use or gift to those you love. Your own honey extraction area. The tote is able to extract all the wax tops out of the honeycombs along the edges. Because honey extraction could become a complicated process buy a bag which is 6 inches deep and 3 to 4 feet long and 18 to 20 inches in width and comes with the top. The ideal place is covered with warmth, making the flow of honey more easily and offers plenty of room to store your equipment.

Extractor

Different types of extractors are available. Smaller extractors typically come with hand-wrench tools, while those with at least six edges are usually electronic. Certain extractors come with plastic tanks, however larger tanks are treated to stop the rust. They use radiating power to throw honey from uncapped cells onto the dividers in the extractor. The honey, at this point, will sink to the bottom, and a valve allows the honey to be poured into a container or tote. Nourishment Grade Storage Containers are an aluminum can with a lid. But, it must be constructed using the highest quality of plastic for nourishment. It is possible to purchase them in 1-gallon, 2-gallon and 5-gallon sizes at an area home improvement store or purchase a pre-owned one from your local bread kitchen.

Hot Knife

You can purchase an electronic cutting blade to remove the cap, or make use of a blade you are using like the serrated bread blade. If you are using bread blades, you need to boil some water, drop the blade in the water until it is hot and then cut off the wax cap that is affixed to those honeycomb cell. In any case, cut off the capping using a virus blade but it's not as easy. Be sure cutting the caps in the tote that is uncapped and save the capping for rendering it into beeswax. The tools are used to remove wax cappings and release the honey. A roller for uncapping is a device that has spikes all over the moving surface. However, the uncapping scratcher is made of metal tines that can penetrate. The choice of which one you choose depends on how little damage you want to do for the tool. The screen will capture most of the wax particles, as well as any other bee components, meaning your honey is ready to be bottled and eaten. Certain beekeepers employ larger screens in the initial run, and, after that, a smaller inspection of a scale channel that is miniaturized prior to packaging to prevent honey crystallization. tiny particles of dust and wax in honey could speed up the process of crystallization however, almost all honey that is thick will eventually develop into a crystal.

1.) Hot blade

2) Uncapping Roller

3.) Extractors with distracting hand-wrench

Step-by-step instructions on how to remove bees from an Super

Smoke boards utilize a synthetic scent that the bees do not like. Find one that is likely to irritate your bees, but will never harm them, like, for instance, Fischer's Bee Quick. Then, it is able to chase them away from honey supers. A smoke board uses an artificial smell that bees aren't fond of. Discover one that could bother your bees but never harm them. It also chases them away from honey supers.

1MAKE A FRAME that is the same dimension length and width honey super but just three to five inches further. Make use of felt or material to cover one side of the.

2SPRAY THE CHEMICAL on the surface, then set it on top of the honey supers, not on their top. In about 5-15 minutes significant part of the bees will be leaving the supers that contain honey.

3PULL THE CAPPED SUPERS out of the hive and put them in plastic containers with tops, or add them to the bee yard, and put them into another structure that is protected from bees and crawlies.

Place the smoke board over the honey after splashing it. Make sure you brush bees gently or, most likely, they will not be in the same way to your. Use your bee brush or lots of long grass to scrub the bees off of each super and every casing separately. When you've got a casing that is free of bees and you are ready to move the outline away from the bee's yard or relocate it to an enclosed box to protect bees from any social gatherings that might occur upon it.

Utilizing Bee Escapes

Bee escapes are one-way valves that can be placed between brood homes and supers of honey you're removing from. It is a valve that bees are able to exit through however they cannot return to the super. The easiest method to make use of these supers is to put some supers of honey on top of the inside spread along with the departure of bees and believe that all the bees will leave down into the chamber of brood. It could take within the range of 15 minutes to three days for the bees quit; however, you'll have a couple of stubborn bees to cross before collecting

Extricating and Filtering Honey

After you've evicted the bees from your honey supers and robbed the honey of their hives, then you'll have to get that honey out of the top casings. It can be an uneasy procedure, but when you've tasted that fresh honey, you'll realise that everything was justified despite the difficulty. There are two kinds edges extractors: disorienting and spread. Most of the smaller hand-turned extractors are not related and therefore that the flat surface of honey's casing is facing the entire extractor. They are able to take just the surface that is the honey's edge. You'll need to turn the casing over to release the other side. Get treated or plastic extractors as they will not corrosion. It is also necessary to buy nourishment grade bearing oil to prevent the heading from slamming. The extractor you choose to use needs be equipped with a honey entry valve that is located near the bottom of your tank.

Removing HONEY

1- SLICE OFF THE WAX CAPPINGS. Make use of your serrated knife to remove the wax capping's from one of the massive cans. This will allow you to access the cells.

2- LOAD THE UNCAPPED FRAMES INTO YOUR EXTRACTOR. Every space that has an enclosure, make an effort to use the same size to adjust edges of the extractor. It is important to keep the honey entrance open when making honey in a manner that the tank is too deep with honey. The casing spinner can stall with this honey.

3. Place a food grade with an sifter or screen across it beneath the entrance. In the course of extraction the wax will begin to block the screen. You can use a spoon to expel the wax, then add it to the capping of the wax that you will slowly strain later.

Utilizing Extracted Honey

TURN THE CRANK ON YOUR EXTRACTOR.

Begin gradually and gradually get moving to the compel, but not as there is as much honey in the cells will be expected. The honey will fall on the partitions of the extractor, and then slide into the base, out of the honey entryway, through the sifter, then in the storage container. Let the honey rest over a period of time before you and then access the basin fixed. The wax particles will rise towards the peak. Make a piece of wrapped saran wrap and place it carefully over the top spot of the honey inside the container. Be careful to lift the plastic upwards and away from the honey and it will remove all the wax out of the honey. This leaves only the honey that is currently being packaged.

One of the greatest benefits of keeping bees is the ability to collect honey for eating, an activity that you can do without any effort. If, however, you do have the possibility of wanting to share your honey with others There are other methods to use. Take a look at Food Safety, possibly you've been told that honey is the most nutritious food because it doesn't cause any harm. This is correct, and it is based on two crucial ingredients:

1.) Honey must be kept in a sealed container, to ensure that it doesn't absorb any additional moisture. It will weaken the honey.

2.) Each when you open your honey holder, moisture could be able to sneak into. Make use of smaller compartments in the event that you don't intend to use honey often. This will minimize the impact of sticky on the honey.

3.) Honey has a pH that is between 3 and 4.5 It is extremely acidic. In addition microorganisms as well as life forms cannot survive in an acidic zone. This is a nightmare for these creatures but is a great thing for you.

4.) Bee catalysts divide honey into two effects hydrogen peroxide and gluconic corrosion and we, as a species, know what hydrogen peroxide does to bacteria. This is why honey has so many health benefits.

Some honey producers over-filter their honey that they package, and even remove the dust from the honey, effectively eliminating the mark. Once the particles have been removed it is difficult to determine the origin of the honey or when it's unadulterated honey. Because a large portion of these particles is dust, which offers advantages, you should to minimize the filtration. Containers made of glass, such as canning bottles or the plastic bears are ideal to store and package honey. Whatever you choose to put in your honey storage be sure to purchase an appropriate cover that seals and seals effectively. There is no need to consume honey that's been destroyed by external contaminants prior to packaging honey. Clean and disinfect the container and cover. To transfer the honey into containers or jugs, you could use the pour ramble, or pipe, or put the honey in tanks or containers with the help of a valve or honey entryway.

Be sure to leave a small air gap between the highest part of your container to the highest. The honey doesn't have to be resting on top of the container while you work it is sifting away. I really love the rectangular glass Moth bottles that make use of plugs to secure. However ensure that you keep the stopper as well as the jug surface clean for an excellent seal. If you intend to sell your honey, it is important to recall some details to give it a name.

1.) It must bear the main name , honey. It could be a combination of the blossom or plant of the plant, like the orange blossom honey.

2.) If the principal fixing you are using is honey then at that time, there is no need to deal with a list of fixings but in the event that you have included other fixings, it is recommended to be able to show them in a typical fixing articulated.

3.) You must also include your contact information as well as the name and complete address of the maker of the honey, the packer, or the merchant (where honey is packaged) in the name that is first.

4.) You need to print the net amount of honey (short the jug) in pounds/ounces as well as the an metric weight in the base third on the label.

Beyond these essential requirements for marking, it is important to consult with the state in which you plan to sell your honey in order to learn about the different laws you must follow in order to sell your honey.

Honey Extraction in a Top-Bar Hive

The fact that bees are kept in a top-bar beehive will mean you don't have purchase an extractor that is specifically designed to remove hives that are surrounded. For collecting and concentrating honey from a top bar hive, and afterward, enjoy the delicious nectar.

You could also cut the honey from the bar in the bee yard. place it in the container, and then place the bar back in the hive, and let the bees wash it. This could trigger an era of looting in your apiary. Therefore, use caution. Bees create brushes when they require they need it (typically brood brushes initially) to the other side of the body of the hive. Then they create honey stockpiling brushes later. You can assist in removing those honeycombs by holding them to an end of your living portion of your top bar beehive. I prefer to keep one honey casing per brood to allow the state to winter However, in the unlikely chance that they have additional honey, you are able to steal it and store on it for your own use.

Chapter 12: What Equipment You Will Need

This chapter you'll find out what equipment is needed to keep bees.

Beekeeping doesn't require a ton of equipment that is specialized, however there are some interesting tools that you can test out. Let's take a review of the basics. After you've completed the first season, you may want to think about adding some tools.

The Hive

There are a variety of hives you can choose from however, they all have the same thing: A box to store your bees. This Langstroth Hive is the most widely used and is a great choice for beginners. In this book, it is this Hive I will be focusing on.

The hive is simple to locate, simple to set up, and simple to maintain. It is available in various sizes and easily purchase the equipment needed to build it yourself online. The advantage of staying with the same design that it's easy to find parts as well as advice the subject, and it's not as costly as different designs.

If you're good at making things yourself, it's easy enough to make your own beehive. The most important thing to keep in mind is that the insides of the hive should be separated by 3/8 inches. This is crucial because it allows for the bees an attractive, even comb. It also allows enough space between combs to ensure that bees don't have to connect combs that aren't.

The comb frames can be removed to assess progress and make sure to ensure that the hive's health is maintained. You can select one of the eight frames or a ten-frame model. The latter is more easy to maneuver since it's less heavy, however the ten frame models permit greater honey production. It is up to you to determine what is most important to you.

The hive is made up of the following components:

* The stand for the hive The stand raises the hive off the ground in order to help it stay dry and increase air circulation. The

elevated position also helps keep the entrance from getting blocked by tall lawns or any other plant. The stand is equipped with an entrance platform for bees to land on.

The bottom board The flooring of your hive. It helps to keep out moisture and makes make the hive more comfortable to live in.

A reduction in the entry point: Imagine an hive in the same way as an actual castle. The more narrow the entrance of a castle and the more secure the castle is. The narrower the entrance, the better. It helps to protect the hive from being attacked by other bees who might be looking for honey. It is also important while the colony is still in the process of settling. If it is hot then you may want to remove the entrance so that the hive can stay cooler.

* The body of the deep-hive: The body is in which the honey the bees need to feed themselves is kept. It is necessary to have two of these, one over the other. The bees use the bottom layer as type of nesting place. On top, they store their honey.

Queen excluder isn't necessary, but it is extremely useful in separating Queen from her supers, where the honey is kept. It is impossible for her to get through the hole into the supers, and therefore cannot lay eggs there. If she did, bees would keep pollen within the supers which could cause the honey to become less clear. When you've gained more knowledge, you'll be able to determine on your own whether you'd like to make use of an excluder, or not.

* Shallow super honey frames: These are the ones which you'll be capable of collecting honey from. This is the area that bees will use up after all the layers below are filled. If this is your first time establishing a hive put them in approximately two months after the time you put the bees into the hive. After the first year the bees are able to enter when the flowers begin to blossom. Supers are designed to be simple to manage. You can also add supers to the beehive by stacking them in case you're looking to collect more honey.

* Frames: Each frame is made of an extra sheet to create a solid foundation for the bees to build the honeycomb. Frames come in a variety of sizes and are able to be picked according to your requirements. This is the foundation. They're thin sheets, made of either plastic or beeswax. They are laid on frames. They create a structure for your bees to create a comb and also build cell structures on the two sides. You can pick a framework made of plastic or beeswax. Plastic is more suitable for its longevity, since beeswax is less durable however, bees might be more reluctant to use plastic. Frames made of beeswax are pre-stamped with hexagonal designs to provide your bees the ideal guidelines for neat and even combs. It is important to insert the foundation in your frames.

The inside cover They are available in a variety of types of wood, from cedar, Masonite and even plastic. The plastic frames typically cost more to purchase, however they won't require replacement them as often as wooden frames. The frames made of wood are durable however they will eventually rot. They are the most fragrant and are the most classic of the alternatives. The use of Masonite can reduce the cost in the beginning but it can end as a waste of money since it will need to be replaced more frequently. Pick a cover that comes with an air vent cut to allow for better ventilation in the hive.

* The cover's outer I was fascinated by beehives as a child. There was something in the small wood "bee huts" (as I often called these) that I loved. The cover that covers the exterior is what makes the beehive unique in design and should be made of a wood like cedar , for the most durability. Doing your best to cut corners on this is not a great option as the wood is totally subject to weather conditions. Plastic is a viable alternative that can last longer than wood, but I cannot claim that I would prefer it. It doesn't have the same style and smell like wood. Make sure you keep the cover in place to stop it from being pulled off by the force of wind.

The Hive's top feeder There are numerous different feeders that are based on the same concept however this one is the most suitable for beginners. It can be utilized in lieu of an internal cover. It has a screen the bottom that the syrup flows through. It is a good option if you're slightly nervous about the bees since it forms an obstacle that separates you from the bees at the feeding period.

Smokers

There are many fancy gadgets that are enjoyable for children to use. This isn't the only one. A quality smoker is an important instrument. It lets you examine your hive and stop the bees from swarming around you while you're doing this.

It isn't easy to keep it burning initially but if you keep trying you'll get knack of it in a short time. The next step to master is the amount of smoke to put out. It is a subject you'll be able to master with time. Always make use of the least amount of smoke you can. You should be able to maintain the bees within the hive, but not too much to exhaust them.

When selecting a smoker, it's worth spending slightly more to get a high-quality one. The design here isn't much as important as the performance - the most effective smokers come with high-quality bellows, and are constructed of stainless steel. You can spend a little more on the best set and you won't need to replace it in the near future.

Hive Tool

The smoker and hive tool should be always along with you whenever you visit your beehive. The hive tool helps you unlock your hive, unwind any pieces that are stuck by wax, and help to remove the frames from their place.

Frame Lifter

This is a different tool employed to lift frames from your beehive. Frame lifters allow you to have a solid grip on the frame in case you need to remove them.

Protective Clothing

The most effective option is to purchase bee suits. This is basically a hoodie which protects you from the stings of bees. It is secured by a strap at the wrists and ankles. The next option is to buy a specially-designed bee jacket and wear long pants that you can tuck in your shoes. If you're not planning to purchase specialized clothing , then wear pants or jeans and tie the bottoms using an elastic band. Select a cotton shirt with sleeves that are long to protect the upper part of your body. It's a good idea to wear the pullover, not an outfit with buttons. Close off your arms using the rubber band or similar.

A Veil

In this particular area, no other thing can accomplish anything. Bees could be gentle or not be. There's no way to know before you arrive at the hive whether something caused them to panic. It is also possible to be doing something that alarms them. It's not worth the risk of the safety of your eyes and face.

Even if the insects aren't hostile, it can be uncomfortable to see bees crawling all over your nose or getting into your ears. Pick a veil with enough protection but is light enough to be used in the summer. It's also helpful to have an extra or two in case you wish to showcase your bees to another person. You'll be happy with your bees as well as what they can accomplish and there will be those interested in them too So be ready.

Gloves

It's easy to imagine that gloves are something you would use regularly, but the issue in beekeeping glove is they're constructed from a heavier material. It is easy to get tired of them as they make it more difficult to grasp objects. They make it difficult to be aware of what you're doing, and you could accidentally injure your bees.

The majority of the time you're not going to require gloves in any way. Remove them at the time:

* You're harvesting honey, and the bees aren't happy to steal their food.

* If you need to move the hive's bodies.

* There is a colony is is more aggressive or it is approaching the final stage this season.

Keeping It Together

Make sure you have a toolbox to keep all your equipment in. Keep your most frequently used tools in it and carry it with you when you visit your beehive. What you'll have to store in it:

* The hive tool you have as well as your frame lifter.

* Rub alcohol on your hands to help remove honey-like bits off your hands after you're finished. (Baby wipes and hand sanitizers are great alternatives in case you aren't a fan of the idea of alcohol.)

Matches to light your smoker. I always have two matches in my toolbox, just in the case.

* A talcum powder that is not scented to dust your hands with so that it is easier to get rid of sticky bits later. Alternately, keep a pair or surgical gloves inside your box to stop your hands from getting soiled.

* A notebook to write notes in , the pencil or pen. It's useful to review the things you've done and observed within your journals. You can record notes to yourself or keep a simple record. This document

can be helpful in the future when it comes to maximising the harvest you have made.

A hammer as well as some nails are good to have in case something goes missing.

* A screwdriver with a sharp blade, and pliers.

A magnifying glass and flashlight. It's not necessary however it can be useful when you want to assess the condition of your hive or when you want to know what happens to bees throughout the day long.

* A larger blade that can be used to break apart the combs which have become fixed to frame edges that make it difficult to remove the frames.

In this chapter, you've learned:

* What components of your hive is composed of.

* A smoker, hive tools, and veils are the most important tools that every beekeeper requires.

* You should restrict your skin that bees can access.

* You don't necessarily require gloves.

* There are additional useful tools you can carry around with you every time you visit your beehive.

It is important to keep records for beekeeping.

In the next section, you'll be taught how to choose the best bees for your needs and how to deal when they arrive.

Chapter 13: Deciding How To

Acquire Your Initial Bee Colony

PURCHASING YOUR FIRST HONEY BEES
It's winter. Although outsiders might think that this is a time of enjoyment for beekeepers, it's actually a hectic, critical time for preparation for the season to come. This is the time when you're putting in your beehive equipment, purchasing materials for your beekeeping, cleaning your honey house and getting ready for the upcoming season. It could be very detrimental later on to not plan ahead. For other beekeepers it's the same. If you're a beginner beekeeper and haven't purchased your first beehive this winter, it's the best moment to make the deposit on the first big bee. But many modern beekeepers don't know the best way to go about this. Here are some guidelines to help you get started:

Try to purchase bees from an beekeeper in your state should you require. This is important because local bees are able to adjust to the conditions of the surrounding environment and have a lower possibility of contracting diseases or pests from outside. But, should you require assistance, there's an emergency contact nearby.

Begin to understand about the information you need to know prior to making your first purchase. You can find clear guidance on the internet and in bookshops for new beekeepers.

Purchase your supplies for your beehive during the wintermonths prior to purchasing your bees. It can help you install the tools for beekeeping as well as to understand the different components that make up the hive that you are buying. Select a reliable beekeeping business that you can purchase equipment and equipment, but preferably one who is an actual beekeeper.

Beware of bee packaging. Bees' packaging can be a nuisance for both the bees as well as the queens. A majority of companies that package bees gather bees from hives that are separated together, and then add the queen outside.

We suggest that you purchase the Nuc hive. A Nuc is a type that is a regular hive that is tiny in dimensions. The bees live in the same hive that is larger, usually containing five frames of honey, larvae and pollen. This gives them a huge break in the early morning since they don't need to head out immediately to search for an abundance of food. The other aspect is that bees that have hive packets are required to fly away to search for nectar and expand to last as long as they can.

In the winter months, make the down payment on your bees prior to the beginning of the season of spring. Bees are sold out fast in many towns If you're waiting for too long, you might not be able to get the bees you want!

PACKAGE BEES - HOW TO BUY HONEY BEES

Package bees are a way you can purchase honey bees in order to start beekeeping and honey production. Beekeeping is about getting bees that live in your beehive! After deciding to begin keeping bees, whether for a leisure activity or as a commercial venture, you bought or built your beehives, and then had them set up in the best place. You've got all the necessary equipment and now it's time to manage those honeybees yourself! To further develop your project, follow the steps below.

BUYING ESTABLISHED HONEYBEE COLONIES

For instance, it's more secure for a local beekeeper to buy honeybee colonies which are well-known. It is safer to purchase two colonies so that you can exchange broods and honey frames in the event that one of the colonies is less robust. A colony of honeybees is made up of between 20 and 60,000 honeybees, one queen active with 10 to 12 combs, larvae, as well as fruits. You can also purchase the 'nucleus' (which is composed of five to seven sets (including fertile queens market, workers broods, and possibly drones). The regulation of the smaller number of bees that reside in either a nucleus, or a "nuc shell' would be more suitable. For beginners it is the most suitable option. The disadvantage of this method for acquiring bees is that this is the most straightforward method to get started. By now, you've got honeycomb and brood, and all the bees that you require. It is possible to purchase colonies and honeybee nuclei on the internet however, it is advised to purchase the bees from reliable local beekeepers and suppliers. There is no doubt that the bees you breed that are in your area are adjusted to the local environment.

HOW TO ORDER PACKAGED BEES

Another option is to buy imported bees to add to your brand new beehive(s) (each package will contain worker bees as well as one queen bee). Honeybee boxes you purchase usually arrive by post in packages ranging from between 2 and 5 pounds. They contain between the 9,000 to 22,000 bees. A three-pound delivery of bees is about 10,000 bees, and one queen that is about the size of the size of a shoebox. There is a protective shield surrounding a box of bees sent via mail, and a sugar cane water in the box to allow the bees to drink during the trip. A enclosure is utilized to identify the queen from other bees. The price of the enclosure plus packaging for an enclosure of three pounds will range between $50 and $70. As opposed to a colony made of ready-made bees the worker bees you purchase in a kit will need feed for a brief time after they are transferred into the beehive (s) prior to when they start to gather pollen and nectar, and then build honeycombs for the food they need. If you are looking to purchase bees for your hive and you are able to do it when it is the best moment. The early part of April is when the bees will arrive , and will be put in the hive, giving them to grow the colony prior to winter arriving. When you place your order as quickly as possiblein January ensures that your bees are scheduled. Did you receive your order of bees from the seller? The bees in fewer numbers will die than if you receive the order via the post. If you get a package of bees via mail, make sure you contact your local post office to notify you once it arrives. The quicker you move the bees to move the better, as it is the more likely they'll be hungry and exhausted.

PUTTING YOUR BEES INTO THE HIVE

It's time for you to put up your bees, now that you have your bees, all the equipment, as well as your safety devices! The entire process can take about two hours from beginning to end. Take your time and carefully step around. You've got everything in a secure position to be ready for deployment.

A STEP-BY-STEP GUIDE FOR INSTALLING YOUR BEES IN YOUR BEEHIVE IS GIVEN HERE.

1. Sprinkle the Bees

Spray the sugar-water mixture on the bees. Utilizing a different container so that the addition of additives doesn't harm the bees. You can also add one-third of the water one portion of sugar. They begin to consume the honey after being treated with sugar water and bees with full bellies are content and peaceful. Also, bees do not have an hive, pollen or queen inside the box they try to conceal. They appear to be quite docile.

2. Spray the Foundation Frames

Pick each frame within the hive body and spray the sugar water on the bases' side. The frames should be covered when you've finished wiping them with a few squirts of sugar water per side is enough.

3. Make use of to use the Hive Tool to take off the Sugar Water Container

This is your very first real bee touch! The secret to bee touch is peace and a slow, steady movement. Bees are just like any other wild animal They are scared of sudden movements. actions can be alarming. If you've watered your bees in sugar water for quite a while and then decide to repeat the process. After opening the canister ensure that you don't lose the queen's cage within the container. In the event that the sugar syrup bottle was removed from the frame, it can be placed over the frames. There are a few bees that will still be hanging onto the container; you can shake them away to the beehive.

4. Take the Queen's Cage with care from the box

Many companies have different strategies to get rid of the queen. Honeybee Genetics has a sealed sugar-paste tunnel, which connects directly to queen. Cut the cap off and scoop a small portion of the paste for the queen to go through it and then be allowed to enter the hive after just a couple of days. It is necessary for the queen to stay in the cage of her hive for a few days to release her pheromones and inform the other bees they have a queen. If she's discharged too quickly, her bees could kill her or disappear. Install a wire inside your cage via an opening and then tie it to one of the frames for the hive. The cage can be removed in three days by removing the cage from the frame to determine whether it's freed. The other companies don't provide brushes. If you purchase from one of these firms take off the cap of the queen's holding tube, and put inside a mini marshmallow. In a couple of days, the queen and other bees need to chew the marshmallow.

5. Shake the bees across the hive.

As you build your queen's beehive, it is time to include her bees in the mix. Take your pack in two hands, and then shake your bees across the beehive. If the bees are stuck around the perimeter, hold them off the ground for few inches before lowering them gently to allow them to move. Continue to toss the box on top of the hive until you have an overwhelming majority of bees left. Set the package in a slant toward the entrance of the hive and let the other bees depart the package for their queen. You don't have to think about bees flying off because they've formed a bond with their queen to become attracted by her in the beehive.

6. Be patient Apply Pollen Cake and Feeder Sugar Syrup. Cover

The majority of bees have left frames within an hour. In addition to the frames the next step is to put the pollen cake, and then place the hive's cover and Telescopic cover on top of the frame. The hive cover is made of made of wood, and typically your telescopic cover will be made of metal. If you decide to leave your hive at the place where the box was then send your bees' sugar syrup through the feeder, then tap it to the bottom. Install a reducer at the door then you're all set! Once you've moved the hive, you need to tape down the cover and door, then transfer the hive with care into your vehicle. The hive should be moved carefully towards the desired spot and then untap the entryway.

HOW TO MAINTAIN YOUR BEES HAPPY AND HEALTHY

A typical beehive comprises a set of crates that are stacked with some intended to contain honey that is processed, and others for hold the brood colony. The upper one that contains honey is often referred to as the honey super, whereas an lower box housing brood colonies is known by numerous names such as brood box deep super or even deep. There could be a flat-screen , also known to be the queen's excluder in between the supers of honey and brood boxes. However, this isn't the case in all beehives. This feature is meant for keeping the queen focussed on reproduction, while the supers that are filled with honey are full of honey thanks to the worker bees. The way to look at the beehive is to smoke it manually and scrape every box until you get to the bottom and then carefully examine the frames within the boxes and observe what you can observe before putting the hive back together.

PREPARATION

Wear your bee suit or sweater and hat so that you can dress to go through an audit. Bring all your smokers and the bee hive equipment. If you'll need to refill feeders, make sure they are ready to be filled. The smoker should be lit and let the bees' to expel delicious, cool smoke. The rules for every beekeeper to follow when checking the hives are as follows.

OPEN THE HIVE

Smoke the front of the hive, in order to disorient the guard bees. Slightly lift the cover, and then steer some smoke puffs under. Allow the cover to slowly descend and allow the smoke to begin to work for about one to two minutes.

FUN FACT

Smoke "calms" the bees, however what it does is sends out an indication that there's an open fire nearby that induces them to eat honey. They do not care about the tall white-skinned animal who pounces on them when they're slurping. When you look around and deciding to take another smoke once you notice their heads lining up in the upper bars.

REMOVE THE OUTER COVER

Take off the cover on the outside of the hive, and then place it on its side carefully and upside down onto the counter. In the event that you own one, pour any smoke that is escaping into the hole inside the cover. Give it a couple of hours for bees respond to smoke.

REMOVE THE INNER COVER

Use your hive tool to cut and tear the interior cover gently. If the inside cover is contaminated with propolis or wax, make use of the hive tool to remove it. Then lower the cover down to the ground over the over the cover. ensure that you do not harm any bees.

REMOVE THE HONEY SUPER

Use your hive equipment to lift the top of the box, honey mega. Take the super off and place it over the top of the cover. The super honey could be thin, medium or deep. If the hive is stocked with super-second honey, simply smoke it and empty the package. If the hive is equipped with an queen excluder that is located under the honey supers take it off and put it aside using the machine in the hive.

Not all colonies can claim mega honey until they are. If yours is in need of one, look in the deep boxes that house the colony.

SMOKE THE SECOND-DEEP BOX

Inhale smoke slowly in the next box of the hive. This is referred to as the second deep. It is one of the two boxes that house the brood colony in the majority of beehives. When you've got three small boxes rather than the two deeper ones, then you need to repeat it at least twice before reaching the box at bottom. The box at the bottom lets you keep going through your inspection.

REMOVE THE SECOND-DEEP BOX

Remove the second layer and gently place it over the honey super layer. This box can be inspected in the future.

REMOVE THE FIRST FRAME

Begin your inspection with the box with the deepest part (bottom). Then, take the first frame, and place it inside a frame holders gently on top of other hive boxes, or on the inside cover, making sure to not hurt any bees. Examine the frames One at each time, with your hive tool, carefully pry every frame free, then pick the frame up and look at it.

Try to locate the queen. It's better if she is identified, however in the event that it's not, it's possible. Check for her lean, long, non-striped belly and the classy circles around her. If you cannot find the queen, then the discovery of eggs is important and indicates that the queen was present within the last up to 3 days.

Find any insect or rodents such as mites, wax moths and foulbrood and so on.

Calculate the number of frames stretched out, and then pack them with brushes that are ready to be used. If seven frames are drawn using the box with the deepest in the middle, then the time is now to put in the second box. After dipping seven frames out of 10 frames in the second cool then add super honey. When the Super Honey is past max Add another.

CHECK FOR LARVAE

The frame examination includes brood-capped and uncapped larvae as well as sperm. The picture below shows a stunning sequence of exposed larvae growing and this is what you're looking to find in your inspection of the beehive.

CHECK FOR EGGS

Finding eggs is the primary part of the beginner's exam of the beehive. However, novices still have trouble finding eggs difficult to find. Eggs resemble tiny rice grains. Two eggs per cell are put in the middle. If you have several eggs in each cell, you've got laid worker bees in your hive . seek out a knowledgeable beekeeper for the possibility of having that. The most efficient way to spot eggs is to have the camera pointed at the sky at about 30 degrees and the sun's bright light reflecting on your head. Make sure to keep it to the left, to ensure that the fabric shadows of your veil doesn't obscure the eggs. Sometimes, it's helpful to wear reading glasses or magnifying mirrors. The camera should be turned in a circular motion until you can see them, and experiment with the angle of the sun and lens. The frame's core on the bottom is usually the most convenient place to clearly mark eggs.

REPLACE THE FRAMES

Put it in the space that was left behind by the frame that you cut, and go through each frame. Then, move each frame to the front of the next one as you remove it gentle, in order to avoid harming the bees. A bee brush or smoke will help get the bees away from the container and away from the ears of the container, which is in the event that they be pinched. Examine the frames that are in their place and, after testing, do not alter the direction. Make use of your hive method to clear space to the front for the first frame. Then, you move the entire set of frames as one unit before you get to with the last frame. Utilizing your hive device, create space on either side of the last and first frames, when you take out the first one so that the frames are evenly distributed within the box.

REPLACE THE SECOND DEEP AND HONEY SUPER

After looking over the first deep box, you can move to the second-deep one and examine the contents, and then return it to the first box. Take off the queen excluder, and take off the honey super, should your hive have one. For this, you need to place it on the back of the hive, with its edgeand then gently "bulldoze" it forward, making sure to be careful not to harm any bees. The beebrush or smoker is a good tool to move the bees out of the way.

REPLACE THE INNER COVER

Utilizing the bulldozer method to put on the outer cover: Begin at one end and gradually slide into the container. Utilize the beebrush or smoker to force bees away when needed.

REPLACE THE OUTER COVER

Replace the cover that covers the hive gently. Then, you can fill your bee journal or bee book to record your observations. Make sure to do it immediately because the exact date and information about the inspection can be lost in the shuffle. Get your jacket off and put the smoker away so it is able to easily extinguish.

Chapter 14: Planting The Right Ones Around

You must be aware of the plants and flowers you should have around your home, so that you can be sure that bees are attracted by them and would like to reproduce.

Additionally, these plants will add some energy to your garden or beekeeping zone. This means that you not just look after the bees, but also help to improve the quality of your surroundings, as well!

Bluebells

Bees are drawn not just to color but also to smell. And they are unable to resist the vibrant bluebells' color! Bluebells are also brimming with nectar that honeybees absolutely enjoy. Even hoverflies and butterflies like bluebells so pollination isn't an issue.

Hellebores

Hellebores are spectacular because they produce a lot of nectar, especially in spring. So, you can ensure that you'll be able to provide honey for the bees.

Rosemary

Rosemary is a favourite herb for humans, and the bees adore them too. The plant is adorned with tiny flowers that can make bees enthralled, and of course, the smell of the plant helps to draw them in and also.

Forget-Me-Not

Another fantastic plant in the season of spring, forget-me nots are a popular choice for bees due to the fact that they have a large amount of nectar which bees require in order to live.

Pussy Willows

The greatest thing about the willows with pussy feet is that they can easily produce flowers. Furthermore, because they are able to produce appealing flowers, bees tend attracted to them, which allows them pollinate plants.

Pulmonaria

Mrs-Moon-image-Bluestone-Perennial.jpeg

Bees are also attracted by tubes-like plants and pulmonaria is certainly one of them.

Thrift

Thrift is famous for its pink clusters which draw bees. They also produce lots of pollen during the spring season and will, again offer enough food to your bees. Therefore, you must to maintain the plants in your yard.

Bugle

Also also known as Aguja, Bugle attracts bees as well as butterflies, and blooms throughout the months of April through May.

Crocuses

Crocuses blossom in early spring and can be an extremely reliable sources of food for bees throughout the winter months. Be sure to keep them in the garden.

Viburnum

Viburnum is a favorite among butterflies and bees because of the attractively-shaped flowers as well as their sweet scent.

Flowering Currant

In addition to beautiful flowers, flowering currant creates the scent that bees would definitely like. They also shine brightly which is the reason bees can be found close to them.

Hawthorn

In the blooming season of May bees will be attracted by the amount of nectar in Hawthorn.

Cherry Trees

Cherry Trees symbolize the arrival of spring and that not only humans are attracted by the trees, but also bees too! They're an excellent source of nectar, and also create a pleasant scent in the air, so both the environment and bees benefit from them.

Chapter 15: All About Honey

How Do Bees Make Honey?

The honey bees keep can be used as an energetic "savings account" for leaner (and cooler) seasons. Bees working out gather nectar, which they return to the hive. There it's converted into honey.

Bees require two types of food. The first is honey made from nectar , the liquid with a sugary taste that can be found in the centers of flowers. Another is made from the anthers in flowers and contain many tiny pollen grains. Similar to the different colors in flowers, they are various types of pollens.

Honey bees are known to collect nectar or pollen (not both). The bee takes nectar out of the flower and store it in her unique honey stomach. Nectar is made up of around 20% water and a few complex sugars. Bees are omnivores with two stomachs: an empty stomach in which honey is stored, as well as their regular stomach. If she's hungry, she'll open an opening inside the nectar sac and some nectar will flow into her stomach, providing an energy source.

The rest is transported back to the honey-making bees' of the hive after her nectar "sacs" are full.

The honey stomach contains nearly 70 mg of nectar. When fully filled, it weighs almost the same as a honey bee. Honey bees must visit between 100 to 1500 flowers in order to full their honey stomachs.

Nectar is given into one of the bees who live within the beehive. They suck the nectar out of the honey bees' stomachs using their mouths. The bees chew the nectar for approximately one-half hour.

At this point enzymes are breaking up the complicated sugars present in nectar, and turning the nectar into simpler sugars. This makes it easier by bees, and less prone to bacterial contamination while it is stored in the beehive.

The bees spread the nectar over the honeycombs until the water evaporates.

This transforms the nectar into a more thick syrup. it is at this point at which the nectar transforms into honey. Bees can speed up the drying process through spreading it out by wing.

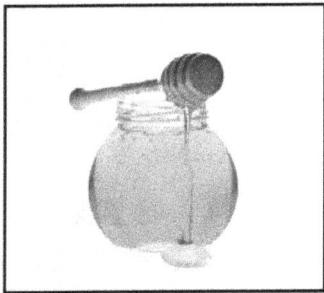

When the honey has reached the right consistency - not too gooey it is put into storage cells, and then sealed with beeswax.

Before embarking on a new foraging adventure The worker bee brushes and scrubs herself. This will ensure that she's working efficiently.

There will be a lot of worker bees out foraging simultaneously. It takes about 300 bees for three weeks to harvest 450g honey. A typical hive is comprised of around 40000 bees.

Although honey bees fly at speeds up to 15 mph, they're not speed freaks. They are adept at making short trips between flowers and need to flail their wings as many as 12,000 times per every minute in order to hold their massive bodies up in the air during the return flight back to the hive.

One worker bee is able to visit up to 22,000 flowers per day. A whole load of nectar on one visit back to the hive is probably result of drinking from 50 to 100 flowers. This kind of volume of activity puts the workers under such stress the workers that they can live just three weeks in the average.

To get a better understanding of the incredible teamwork involved in the production of honey within a beehive A single worker bee can never make more than one-third of one teaspoon of honey over his entire life!

A complete colony, comprised of up to 60 000 bees, has the potential to yield more than 200 lbs. or 90.72 kilograms of honey every year. That's one hell assembly line working!

Honey bees produce huge amounts of honey that supply the entire colony with food during the summer months and also to store food for the winter.

Why Do Bees Make Honey?

Honey bees are distinct in the sense that they winter as colonies. This is the major distinction between honey bees versus other kinds of bees. Food storage, through honey is what helps the honey bee to endure the cold winter months and sustain the colony.

This is similar to bumblebees who have a small colony of about 100 bees. They also collect pollen and nectar as honey bees, and they keep excess nectar and pollen (which transforms into honey) inside a tiny colonial nest.

However, bumblebees do not create a large honeycomb, and they don't cluster in the same way as honeybees. They prefer to let the nest fall and disappear in autumn.

Only the newly pregnant queen bumble bee will be able to survive the winter (either sleeping alone or with some of them). Then, in the beginning of spring all queens will begin to construct an entirely new nest and will be the only egg layer. the sole egg layer, and going out on her own in the early days of her new nest.

This means that there's no need to store the huge honey storage that is available during winter.

Honey bee colonies do not hibernate , but is busy and groups together to keep warm throughout the cold winter. This requires lots of food, which is the reason for the huge collection and storage of honey during winter and spring.

Even though a colony needs around 20-30 lbs. of honey to last through the typical winter, bees will collect larger amounts if there is ample storage space. This is the ideal situation for beekeepers.

Bees have used the same techniques to make honey for over one hundred and fifty million years.

Beeswax

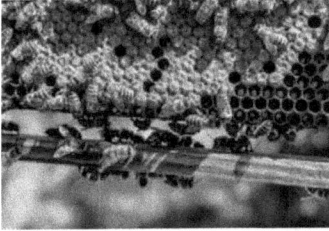

The bees that are the smallest cluster in large numbers to raise their body temperature. Eight pairs of glands located on the lower part of the worker bee's abdomen create tiny slivers wax. These wax drops are able to harden when exposed to air. The wax is removed from or around the beehive.

The workers put the flakes in their mouths before allowing them to soften into the material that they can make into the honeycomb.

The bees that work in the hive gather the beeswax, and use it to build cells to store honey and provide protection of pupal and larvae in the beehive. In actual fact, the comb which honey is kept is composed of hexagonal cells made of beeswax. In these chambers that pollinators also brood.

In order for the wax-making bees produce wax, the optimal temperature of the hive should be in the range of 33 to 36 Celsius.

When it's new it's mild yellow. It will darken over time to a rich gold before it eventually turns brown. Chemically, beeswax consists of the asters of fat acids, as well as a variety of long-chain alcohols.

Beeswax is used in a variety of ways for candle making, from traditional as a component in lip balms, and even as an oil coating for certain medicines.

Beeswax is also extensively utilized in the food industry. It is used to create a glaze (to keep water from leaking out or for protecting the exterior of certain fruits). It's also employed for an external layer of cheese. It creates an airtight seal. airtight. This keeps mold from growing. It can also be used as a sweetener , and natural chewing gum typically uses beeswax.

The varnish is additionally used in production of electrical components. It also serves as an ingredient in a variety of varnish.

Beeswax is a food item, however, for us, it is extremely low in nutritional value. This is due to us finding the process difficult. Certain birds can consume honeywax (such for example, honeyguides) and beeswax is a major component of a larvae's wax diet.

The best estimates of the most productive hives put the production of beeswax at around 24-30 pounds. 11-14kg of beeswax to 1 lb. or 0.45 for honey.

How Much Honey?

A colony of honeybees can yield more than 60 pounds. (27 kilograms of honey) during a peak season. A typical beehive could produce around 25 pounds. (11 kilograms) of honey that is surplus.

Bees travel 55,000 miles in order to produce one 1 pound of honey. This is the equivalent of 2.2 times the distance around the globe.

A colony that is strong will produce between 2 and 3 multiple times as much honey they need. This implies that the bees will not want the honey used from the beekeeper.

In autumn the beekeeper could give the bees sugar syrup. It isn't always required however it can offset any loss incurred by the beekeeper consuming the honey.

Different Types of Honey

There are a variety of honey. Some are clear and flowing, while others are hard and opaque. The kind of flowers and foliage available to bees determines the kind of honey they produce.

Plants like oil seed rape yield large amounts of honey that set extremely hard, so that bees are unable to utilize it during colder months. Flowers in the garden can produce transparent liquid honey.

If the beekeeper finds the beehive outside of the range of a number of sources for flowers, the bees will produce an uni-floral honey.

Examples of monofloral honey are pure clover, or orange blossoms. It can be challenging to obtain - many beekeepers make honey as made from a mix of flowers.

Chapter 16: Dysentery & Brood

Diseases

Dysentery

A lot of the issues we have described thus far could be misinterpreted as Dysentery. It is a condition that result from bees being unable to clean the hive properly over a long time. Like we said, it could be caused by the more colder temperatures. It could be due to bees that are under high levels of stress. When they're in a state of weakness due to health issues, it can cause the conditions that lead to Dysentery.

Treatment and Prevention

If you permit unsanitary conditions to flourish within the beehive, two outcomes will happen. Or, the remaining bees will leave the hive, or they will die within the hive. If you store the hives in a ventilated area during the winter months it is important to ensure that you get rid that honey. Replace it by high fructose corn syrup or sugar water. These substances allow bees to live, but they do not have the indigestible matter which they have to expel from their body.

Chilled Brood

While chilled brood isn't an illness but it should be noted here. It's the result of bees being improperly managed by the beekeeper whether or not they intend to. Sometimes, it may result from using a heavy concentration of pesticide which kills a significant portion of the adult bees within the beehive. A abrupt drop in temperature during the spring could also be the trigger.

Thus one of the main goals of a good beekeeper is to come up with methods to ensure that the brood is cool enough to be able to withstand any conditions. Inspecting the hive before opening it or collect honey may cause the brood to cool therefore, make sure to consider the temperature and conditions into consideration prior to engaging in these activities.

Prevention

It is best to open the hive during warm weather and try to open it during the hottest times in the morning. Be sure that you have the appropriate tools at hand and you've got the proper procedures in the right place. This lets you inspect the beehive within the shortest duration of time.

Brood Diseases

There are two recognized infections which have been proven to be responsible for the loss of huge numbers of honeybees, especially within the United States. These are:

American Foul Brood

European Foul Brood

They could also destroy whole colonies of bees which you thought were healthy and flourishing. The adult bees don't suffer any harm, however, these issues stop the development of enough bees' young in order for the colony to live.

American Foul Brood (AFB)

It is believed that around 3 percent of all honey bee colonies that exist in the United States right now are affected by the AFB.

It's a bacteria that generates millions of tiny spores that can infect larvae. They don't die in the presence of high temperatures or chemical substances.

When the infection spreads about, it will cause the larvae and cause them to end up dying. The infection will lead to the larvae becoming brown in color. The remains of the larvae will cause an unpleasant odor emanating from the beehive. There are a variety of signs of AFB to look out for. These include:

A pepper spot pattern on a brood

Dark colors of the caps of the cells

Deadly remains of larvae

Dark scales are difficult to remove from cells

Treatment/ Prevention

The most common misconception among beekeepers concerning AFB is that they're typically responsible for transmitting between hives to the next. Clean your equipment prior to applying it to other hive to stop AFB infection from becoming the process of being transmitted. The bees all end up dying in the colony and there's not much one can do in order to aid them. To reduce the chance spread of the illness it is suggested to burn the adult bees, the brood combs, and finally sterilize the honeybee hives.

European Foul Brood (EFB)

EFB is caused by bacteria of a specific kind that can be a major issue since it usually occurs when colonies are growing in size. The death that is likely to occur is likely to occur in the larvae stage however all bees are susceptible. If the issue persists and grows, the whole colony may die.

Died larvae can appear white or brown. They could have a foul scent due to secondary bacteria that are linked with EFB. There are a variety of signs the beekeeper should be looking out for that could indicate EFB exists. These include:

Uneven brood patterns

The larvae are brown or white.

Sour odor

Prevention/Treatment

EFB is usually passed on through contamination of equipment or containers, which is why the beekeeper needs to be vigilant with these things. The spring season is the most frequent time it develops, however being vigilant throughout the year is crucial. Every piece of equipment must be cleaned prior to being used on another colony in order to ensure that harmful elements don't get transmitted between colonies to another.

EFB is usually thought due to stress in the colony. However, this is just one aspect that influence it. The usage of a chemical named Oxytetracycline could help, however the honey that is present in colonies will need to be removed. There is too much of a chance that it could be impacted by the chemical residue to consume it.

Chapter 17: Problems Associated

With Beekeeping

In the last period of time, the public activism on natural issues has grown. People began to react seriously to the reality of the chemical concoction as well as the physical destruction of certain aspects of their lives. Additionally, individuals do not generally have a clear picture of the truths and limitations of ecological harm. Usually, this is due to management or specialized issues.

The presence of toxins in the family affects the overall effectiveness and usefulness of honey bees, their disease resistance, and their ability to survive winter in peace. The rapid and massive response of honeybees to natural debasement within the pollination zone activity is a type of equipment to ensure the unshakeable effectiveness of organic control over the surrounding environment. What are the main factors that have an antagonistic impact on the lives of honey bees? From our point of view they can be classified into four major categories:

The primary group, the human-made pollution of mechanical emission - hazardous substantial metals and metalloids, chemical radionuclides and so on. The solution to this problem is possible only at a top-level authority by ensuring a strict operational order and an intense city focus on administrationin the event that there should occur an event that demonstrates the closeness and effectiveness of the unique filtering and removal of modern pollutants. Their destruction isn't by any stretch of the imagination possible without a comprehensive and vast approach.

The second group is the toxic contamination of entomophilous plant species by pesticides. We need a legal system of connection to beekeepers, the Plant Protection Service and beekeepers. Many times, ranchers pollute the woods, fields, and plantations during the rapid collection of nectar by honey bees without prior alerting the beekeepers nearby. The result is a massive destruction and the death of honeybees that fly and the survivors often have nectar and dustthat has been contaminated with pesticides. These when winter arrives, it can result in the deaths of entire families. Although it is true that the use of pesticides can be effective against insects, but at the same time it's unclear what the homestead's management aren't considering a different factor that can boost the efficiency of horticulture, which is pollination. When using chemicals, people aren't always demonstrating the common awareness of the issue, do not take steps to determine the innovation or the timing of defense actions. This issue is obviously a matter of debate be addressed on the basis of local group through the individual warnings to beekeepers regarding the medications that are available in farmland.

The third group is the results of various concoctions veterinarian medications, which are used to treat honey bees. It is now believed that the best method of treatment and counteractive treatment for honey bee-related ailments is chemotherapy. There are many ways to combat infection-causing agents that include the use of various plants within their natural structure or in the form of imbuements, decoctions as well as other devices.

Problems arise mainly due to the manner in which, at times beekeepers don't pay attention to the recommendations of specialists or deliberately harm their own bees, in search of the possibility of a significant restorative effect. In addition, in some regions, people often offer medicines that have not been tested at specific research institutes or not certified by the way recommended and without an official "Registration Certificate". In this case they can disrupt the balance between restorative benefits and the health benefits of the use of chemical. Therefore, beekeepers themselves cause natural problems, by delivering poor quality honey products and then they end up in the food supplies of honey bees in winter and reducing their strength to the point of death, or placed in the food of people who are waiting for therapeutic and nutritional properties, preparing to boost health. Unregulated use of medicines could lead to the accumulation of substances that are not approved in honey bee-related products. They could cause severe damage, particularly if their concentration in the honey bee's brush reaches a point that is dangerous for the brood. This problem can be resolved.

The fourth group is the interplay of genetically altered plants. In the last 80 years of the in the XX century, hereditarily altered plants have been discovered in the horticulture industry (the primary species was the tobacco). There is no conclusive conclusion regarding the benefits or harms of these plants for animals as well as humans. Alongside the transgenic plants, there is the issue of honey bee products. As per EU Control 49/2000 food items that contains less than 1% of hereditarily altered fixings should not be concerned to test. Honey has 0.1 percent dust and doesn't need to be tested to determine the proximity of transgenic products. However it does not take into account the possibility of them being removed from the nectar. It is important to consider ambrosia and dust for the potential of transgenes particularly in the areas that utilize transgenic plants dust, which is used to boost the health of people, and honey bees and honey bees. We must study the impact of transgenic substances on imperial jam, propolis honey bee venom, imperial jam, and beeswax which is commonly used in the form of medicine. Experience from around the world has shown that ranches often sow seeds of transgenic shabby grub corn (maize) and is not intended to be used for sustenance products. In addition, transgenic plants rely in the direction of the conceptual qualities that queen bees possess. Perhaps that's why there has been a drastic reduction in the amount of honey bees in recent times as they can't be imagined, because of the the primary source of their entry into the world - the queen is losing its capacity to lay eggs.

DURING WINTER:

COMMON WINTER BEEKEEPING PROBLEMS:

The time to consider preventing and treating diseases is an ideal time to consider the dangers that could be for the honey bees from bugs and diseases, one must keep an eye open for them continuously. The Varroa vermin is a tiny animal, with the dimensions of a pinhead, which attacks the hive, grows in the brood and lives in the thorax area of the honey bee. It is believed that Asiatic honey bees over the years, have displayed amazing adaptation capabilities to the parasite and have been able to survive without the bug within its honey bee hive. But, if the varroa bug enters the home of the European honey bees, it will have an utterly devastating and destructive impact. Every beekeeper must be alert to ensure that varroa is kept out of their honey hives. Hives have been renovated to aid in this regard, and have floor sections that cross section through which the mite can fall if it is knocked off the honey bees.

Unable to move more into the hive the insects can be collected on a tray under the lattice floor, with the goal of helping remove the beehive. The small chestnut sparkling creatures are dazzling, like tiny pinhead-sized conkers, in all the trash in the hive that have not been able to get rid of. There are many ways to kill them , and aid the honey bees in their efforts to get them out of the beehive. It is possible to kill them by drugs that are based on substances. You could clean honey bee frames with sugar that has been finely ground and which will eliminate the bees and make it easier for honey bees to smash them away. There are a variety of infections which honey bees are likely to. Nosema is one of them. Its loose bowels as adverse effects can be observed in the outside of the hives, as honey bees contaminate the hive as well as the arrival board. There are many deadly diseases that affect brood, also known as European and American "foul broods," but they do not have limitations to the national level. They are so severe enough to warrant in

169

Britain you must inform your honey bee administration auditor if you suspect they could decimate your colony. Once the honey harvest has ended and the honey harvest is over, it's time to treat any illness (so that you don't have any medicines ending into the honey) and ensure that your colonies are as strong as is possible ahead of the winter cold.

PREDATORS It's the perfect time to guard your hives from predators that are larger. Woodpeckers, wasps mouse, foxes, foxes and badgers - all ready to devour honey bees as well as your honey as incredibly nutritious sources for winter foods. Woodpeckers are winter bugs. In January and February , when the ground has solidified the beehive can serve as an easy source of nutritious insects. Urban foxes are more of a problem regularly and willfully wandering around an hive, and then thumping on uneven ones. The warning signs must be put in certain areas strategically close to the hives belonging to honey bees to inform intruders not to enter the honey bee hives. A smaller, but no less of an urban nuisance can be the mouse in your home seeking a warm and comfortable place to rest. They're as much of problem in the beehive as they are in a home, and the best method to deal with their presence is to prevent gaps that they might get through. In the hive , that means the passageway, which needs be reduced to as small a space as it is possible to expect in winter, which makes it easier for honey bees to guard and more difficult for mice to smash through. Most problematic is the wasps. Like all honeybees except honey bees and honey bee colonies, provinces of wasps do not survive winter. After the queen is rested, there's nothing for the other members of the state to do. They stay close by in groups, looking for whatever they can find to consume. The more sweet, the more delicious. Therefore, they'll take your picnics should they smell honey, they'll attack your honey bees. Wasps are able to destroy an entire hive over the course of several days. Typically, when you are relaxing towards the closing of the season, and you decide to go away on the weekends, wasps will come and wipe out your beehive. They'll feast on everything, starting with honey, and the honey bees, too. The honey bees will try their best to guard against the predators. A little hive entrance can assist as will wasp traps easily made from old pop bottles. It is possible to watch honey bees really battling wasps as they move around the hive grappling with each other in the sand. Offer them any assistance you can.

Losing your honey bees is very heartbreaking loss of honeybees. Whatever you do to protect your management and cultivation it is not likely that all or your colonies will last. It's been the demise of settlements which has caused the rise of enthusiasm for honeybees and beekeeping and has brought a number of new beekeepers to the side curiosity. The causes of province breakdown are many, but there are those that are able to be kept in check. If you can keep your honey bees well-fed throughout the winter months, and guard them from pests and diseases chances are that they'll make it through until the spring. In the summer and spring period, when the days are shorter and the temperatures fall the hives, shackled in place and protected from predators, no matter how secure they appear to be, prove to be much more difficult to manage. However it's still possible to keep an eye on the movement of the honey bees that are in groups that are winter-bound. It is possible to listen to the bees by pressing your ear towards the hive. You can feel the warmth leaving the hive and you can observe tiny pieces of wax dropping into the lattice on the floor of the hive, when the state caps are removed. They put the honey away. The picture of the wax fragments placed on the plates beneath the floor also gives the impression of how vast the settlement is as well as where it is arranged. Additionally, you can lift the hive for stores, specifically lifting the hive from one side to see its weight, and consequently the amount of honey that the honey bees are left to eat. If you're bolstering it with fondant, a quick inspection of on the outside of your hive can inform you if the honey bees have consumed all the sugar in the sugar sack. In the winter's gloom the bees don't penetrate too deeply into the hive. There is nothing to do for the colony other than to combat mice and woodpeckers in order to keep it dry and ensure that it is stocked with enough food. The season is now over. It's time to clean out and prepare for a long and tiring year in the coming years. The months are going to be cold and long and it might seem like you'll never get your honey bees ever again. However, someday the willows will be sprouting and the honey bees that are scrounging will be out to collect the first dust for the coming year. The new season has begun.

Chapter 18: How To Market Your Honey

Some of you were fortunate enough to get your own honey . If you were fortunate enough to be lucky, you may have had additional honey that you can sell. The prospect of selling your own honey can give beekeepers a great feeling of pride. Based on the place where the bees hunt, each container of honey from various beehives is distinctive in flavor and the color.

There are several methods to collect honey. You can extract from the comb, strain it, and then bottle it. You can also cut pieces from the comb that is filled with honey. Whatever you choose, there's an interest in both! If you want to make honey for sale, you must meet certain legal requirements established by the regulatory bodies of the United States. It's good to be aware of them, particularly for helping returning customers come back and repeatedly!

Any honey that is to be sold must have an easy-to-read and clear label. The first thing that needs to be printed at the top of the packaging should be the words "Honey". If your honeybees go foraging wherever they choose, simply mark the honey with "Wildflower Honey". If your bees pollinate a specific crop, like clover, cranberries, or the like, then you could declare it so.

Then, you must mention your honey's weight, not including the packaging. This should be included within the third of your label . You should also clearly show each ounce and grams. One ounce is 28.35 grams, and one pounds (16 inches) amounts to 453.59 grams.

Additionally, it is important to add your address, name as well as your phone address on your front of the panel and also on the back. Although it's not required however, I typically include a note on the back of the box to remind people that babies younger than 12 months shouldn't consume honey because of the possibility from Infant Botulism. I also like to add the season and year also. Honey to me tastes different from fine wines!

In the end, based what amount you have harvested Give it to family and friends first. If you're in a position to buy enough contact local businesses selling gifts or food products. I'm sure they'd be delighted to give your honey to sell.

Curbside Honey Sales

The roadside stand is just as common a site in rural America as the lemonade stands is in cities. When a garden produces an excess of produce farmers pull out an old table, load it with baskets of vegetables as well as fruit, then put an open check before it. The exact same stand along the roadside you employ to sell your additional produce and also vegetables can also be used to market your surplus honey that you've collected from your honey beehive.

When you're preparing to set up your roadside stand, ensure that you've got an indicator that indicates that you will be offering honey. The sign should be clear. The letters should be clearly written using paint or ink to create the contrasts to the backgrounds of your sign. The sign must be large enough for people who drive by your home should be able to examine the sign while they pass by the roadside sign. Make sure that the sign is easily visible from the road. Make sure that the design is simple The sign that you're advertising your stand on the road isn't the ideal place to test your creativity. People who fill their cars trying to look at an interminable, tired smirk on a piece of cardboard typically aren't great buyers. Do not be cute and sketch a picture of a honeybee onto your signage, people could not understand and believe you're advising the public of an imminent attack.

Before you put out your honey, you should take an additional look at it. Make sure that the honey is stretched. You shouldn't find any sort particles of dust, lumps, wax, or other foreign substances within the honey. Spend a few minutes and scrub the area around the container to a clean, wet cloth, and remove any the traces of spills. Be sure the container is totally dry prior to exposing the container to dust.

It is best to offer your customers a variety of types of honey. Offer them the chance to buy containers of honey you took from your honeycombs. You can also encourage them to purchase an old honey comb that retains the honey in the connects made of wax.

Give your clients a selection of sizes of containers filled with honey. A few people shy away purchasing large quantities of honey because they fear that the honey might grow before they've had the chance to use it.

Don't be afraid to market other fruits and vegetables alongside your honey. Don't be afraid to offer cut flowers, lovely corn as well as banana peppers. Diverse foods, with various colors can give your stand on the road an exciting tasting.

It is best to set up your roadside plant in a shaded part of your backyard. The shade can make your product appear fresher , and it will also encourage customers to stick about the item. If you notice that your product is beginning to appear old, you can change the color.

If you're selling veggies, do not worry about having to sprinkle the vegetables with a spritz of water. The water will look more fresh in the event that they are sprinkled frequently.

You are welcome to join as customers. They are likely to return to a stand along the road provided the owner is pleasant and welcoming.

Conclusion

Each component of beekeeping is essential, but you should be aware of four essential elements in keeping bees when you're an aspiring beginner.

Bees are always in need of flowers

The warmer the hives, the more bees.

Try as many as feasible to eliminate parasites and increase the efficiency of your bees.

Try as many as you can to keep bees free of toxins.

Do your best to stay clear of pollens and nectars that are toxic, and, if you can make sure to keep your bees out from areas where pesticides and other chemicals from plants are used. If you can, think about using mice-resistant bees which are more efficient and are able to withstand the common diseases that are caused by rodents. They are Varroa mite is one the most common mice species that has the ability to transmit viral infections to mice.

Do your best to ensure that your bees are close to areas with plenty of flowers throughout the seasons. To ensure that you have the most protein-rich winter bees are best to store enough pollen during the fall and summer seasons. As much as you can to eliminate parasites by cultivating only bees that are regionally adaptable as they are able to withstand certain parasitic illnesses, unlike the bees from other countries that are difficult to adapt to different surroundings.

Bees need to be kept in boxes that are tight in order to ensure they are warm. They must be cautiously introduced to new environments without risking their health. In the case of clusters, raising bees will aid spread warmth and make them more active and producing more honey.